Axel F. Cronstedt

An Essay Towards a System of Mineralogy

Axel F. Cronstedt

An Essay Towards a System of Mineralogy

ISBN/EAN: 9783744790123

Printed in Europe, USA, Canada, Australia, Japan

Cover: Foto ©berggeist007 / pixelio.de

More available books at **www.hansebooks.com**

AN

ESSAY

TOWARDS A

SYSTEM

OF

MINERALOGY.

A N

E S S A Y

TOWARDS A

S Y S T E M

OF

MINERALOGY:

BY

AXEL FREDRIC CRONSTEDT.

Tranſlated from the Original Swediſh, with NOTES,

By GUSTAV VON ENGESTROM.

TO WHICH IS ADDED, A

Treatiſe on the Pocket-Laboratory,

CONTAINING

An Eaſy Method, uſed by the AUTHOR, for Trying MINERAL BODIES,

WRITTEN BY THE TRANSLATOR.

The Whole Reviſed and Correƈted, with ſome Additional Notes,

By EMANUEL MENDES DA COSTA.

L O N D O N:

Printed for EDWARD and CHARLES DILLY, in the Poultry.

M DCC LXX.

Mr. DA COSTA's

PREFACE.

I Should hardly have troubled the Reader with any Preface, had not the title page mention.d a Tranflation by Mr. Engeftrom, revifed and corrected by me ; which neceffarily demands fome explanation, in juftice to us both. It is as fo lows:

. Mr. Engeftrom, a Swedifh gentleman of great merit and learning, particularly in the fcience of Mineralogy, amufed himfelf with tranflating Mr. Cronftedt's celebrated *Syftem of Mineralogy*.

On Mr. Engeftrom's return to Sweden, this tranflation became the property of Meffieurs Dilly, by whom I was employed to revife and correct it; as it could not be expected that Mr. Engeftrom, being a foreigner, was capable of giving a correct tranflation in regard to language, or to a proper application of fcientifical names.

I have therefore carefully corrected it in thofe particulars, and collated it with the German edition printed at Copenhagen in 1760 ; and the notes which occurred to me I have marked with D. C. to diftinguifh them from thofe of Mr. Engeftrom, marked E.

THE

THE

TRANSLATOR's

PREFACE.

THIS Eſſay was publiſhed in Swediſh in the year 1758, by the Author, who in the beginning choſe to be anonymous, for reaſons he has given in his Preface: He could not, however, remain long concealed, but was ſoon diſcovered to be the learned nobleman Axel Fredric Cronſtedt. I now give a tranſlation of it, to comply with the deſire of ſeveral of my friends here in England. I ſhall not attempt to amuſe the Publick in favour of this work, ſince it ſpeaks ſo well for itſelf, and has been almoſt generally adopted wherever known.

The univerſal applauſe, and the favourable reception it met with in Sweden, made it ſoon known in Norway and Denmark. In the year 1760 it was tranſlated into German, and was
equally

equally approved in Germany; nor, indeed, has it been unknown to the learned in England; for the ingenious and celebrated Dr. Lewis has mentioned it with praiſe in the ſecond part of his Philoſophical Commerce of Arts lately publiſhed.

As a foreigner I ſhould make an excuſe for the tranſlation, it not being ſo elegant as it ought to be; however, I flatter myſelf I ſhall not be too ſeverely cenſured on that account, ſince it is a known truth, that originals always loſe ſomething of their beauty by being tranſlated: I therefore think it better to prefer the true meaning of authors to the elegancy of ſtile, particularly in ſcientifical works; and I am in this reſpect more able to give the Public ſatisfaction, as I have had the happineſs to be a diſciple of the author himſelf.

That the mineralogical terms might be more generally underſtood, I have added the Swediſh and German names of the mineral bodies to their Engliſh and Latin names; except when to avoid tautology I have ſometimes left out the German as nowiſe different from the Swediſh names. New diſcoveries being daily made in this ſcience, I have alſo added ſome notes of my own, concerning ſuch things of which I am perfectly convinced, leaving ſeveral others to be further examined and tried.

I was in hopes to have ſeen a ſecond edition of this excellent work improved and augmented by the author himſelf; he having,

ever

ever fince the firft publication of it, been conftantly employed in making further enquiries
and difcoveries in this fcience : He had even
actually made fome collections towards it, of
which, however, the literary world is likely to
be unfortunately deprived, as he lately died in
the fortieth year of his age, before he had
time to revife and put his new obfervations in
due order.

THE

THE

A U T H O R's

P R E F A C E.

AS former ages principally encouraged philological and antiquarian enquiries; this prefent age, at leaft in Sweden, favours the ftudy of Natural Hiftory.

Such changes muft be afcribed to geniufes, who underftand how to make thofe fciences, which they have chofe for their principal ftudy, agreeable to the Public; and which Sciences, being in general ufeful to the whole community, every individual thereof can reap fome advantage from it, and thereby gratify that felf-love implanted in the breaft of all mankind in the purfuit of them.

When the pride of a nation is flattered with the vain glory of being of great antiquity, the author of fuch an opinion is always favoured; and every little circumftance conducive to further confirm it, is carefully recollected and noted. Thus when the fubjects of the creation are re-

prefented

prefented to us in a manner which affifts our memory, and renders our conceptions of them eafier, we aim at earneftly adopting the noble improvement, and, in order to be the more efteemed, we likewife always confer praife on the author.

As long as the author adheres to his fyftem, and does not alter it, but only illuftrates it from time to time with fome additional obfervations, we are not only well fatisfied with him, but alfo often become his faithful affiftants. But if he, convinced of the impropriety of his method from its very principles, rejects it, and prefents us with another new and entirely different; what will then be the refult? Or what is likely to happen if this is attempted by a perfon who is unknown, and not artful enough to feize on the advantages of our paffions?

For my part, I am apt to believe, that in the former cafe, the prefent general tafte might be fomewhat leffened without any lofs to the fcience itfelf; becaufe among the great number that love Natural Hiftory, there are always fome who embrace it when free from errors, and others, who are only fond of new reafonings and conclufions, merely becaufe they are fo.

Thefe latter are even of fervice, and their party will certainly increafe in length of time.

From this perfuafion I have ventured to publifh this Effay for treating Mineralogy in a fyftematical manner; a ftudy to which I have with fo much pleafure applied myfelf. It is not done from the defire of novelty; and ftill
lefs

lefs from contempt of thofe fyftems, which
Swedifh gentlemen in particular, very deferved-
ly, though chiefly on the fame principles, have
heretofore generally purfued.

I have thought proper to conceal my name *,
to prevent any conftraint on myfelf or others,
and with a view to be at a greater liberty to
amend the Syftem, whenever I fhall be convinced there is a neceffity for fo doing, either
by my own experience, or by the obfervations
of others : For I flatter myfeif that this work
will not pafs unnoticed by men of letters ; and,
as it is only an Effay, it ought, according to
an eftablifhed law amongft authors, to be
fheltered from too fevere cenfures.

I wifh that the mineralifts themfelves would
examine and compare all that has been hitherto
done in this fcience ; they would then find the
reafon which has induced me to deviate from
the received fyftems, and to propofe another
founded upon my own, as well as upon the
difcoveries of others. But as this comparifon is
not in the power of every one to make, I think
it neceffary briefly to repeat here the changes
which this fcience has undergone.

The firft writers on Natural Hiftory found
fo great a number of unknown bodies before
them, that their curiofity and time would not
allow them to do more than to defcribe them
by their mere external appearances, and to collect the names by which they were known to
the natives of the countries where they were

* Mr. Cronftedt never put his name to this Effay. D. C.

found,

found. But as every country had a different name for thefe bodies, they often gained more names than there were real fpecies, and even fometimes the very reverfe happened; this occafioned a confufion, which in the beginning was excufable, but in length of time could not fail of being an obftacle to the progrefs of the fcience, and its application in common life.

To remove and alter thefe inconveniences, they have in later and more enlightened times endeavoured to fix proper names to the fubjects of the mineral kingdom, according to their external marks, as in regard to Figure, Colour, and Hardnefs; but thefe characters afterwards having been found not fufficient, it was necef-fary to difcover others more folid by the refult of chemical experiments, which added to the former ones would make a complete fyftem. Hiarne and Bromell were, as far as I know, the firft who founded any mineral fyftem upon chemical principles. However, they were only the projectors of this manner of proceeding; and to them we owe the three known divifions of the moft fimple mineral bodies; viz. the *Calcarei, Vitrefcentes, et Apyri*. This fyftem was afterwards adopted by Dr. Linnæus, who, as a very fkilful perfon in the other two kingdoms of nature, ought not to have omitted the third when he publifhed his Syftema Naturæ. Browal, bifhop of Abo, a prelate of great learning, had an opportunity of altering and improving Linneus's method in a manufcript, which Dr. Wallerius has fince made public in his Mineralogy,

logy, with some alterations of his own ; however, the principal foundation remained the
same in all, or according to Bromell's method,
which he had publifhed in a fmall book, enti-
tled *Indications for the fearching for Minerals :*
Untii Mr. Pott, a chemift by profeffion, and
confequently inclined rather to believe the effects of his experiments, than the external appearances alone, proceeded farther than was
cuftomary before his time, in the affaying of
ftones by fire, and afterwards publifhed his acquired knowledge by the title of *Lithogeognefia.*
From this book the faid author received confiderable honour, becaufe the true advantage of
his refearches began to appear : Miners and
other manufacturers were by it able to determine the reafon of certain effects, which they
before either did not obferve, or wilfully concealed, to avoid the cenfure of being ignorant,
if they advanced fuch things as real truths,
which, according to modern fyftems were regarded as contradictory and abfurd. Mr. Wol-
terfdorff, a difciple of Mr. Pott, then begun
immediately to form an entire mineral fyftem,
founded upon chemical experiments ; but his
mafter did not approve of it, ftill infifting that
materials were yet wanting for the purpofe ;
and that *every* mineral body ought firft to be
examined and tried with the fame care that he
had tried and examined the moft fimple of
them ; to wit, the Earths and Stones.

Such was, according to the idea I had of it,
the ftate of Mineralogy, when I, touched by
the

the difficulties which beginners laboured under, undertook to put my scattered thoughts in this order. Naturalists agreed with me, in thinking the barrier, which had a long while been defended with such ardour, was now beaten down, and that it was neceſſary another ſhould be erected in its ſtead, as good as could be procured, until a perfect one might be in time diſcovered.

Such an enterprize it was thought would promote this deſirable end, when on one ſide I reflected upon the paſſion which our learned have for diſputing, and on the other part conſidered the gracious reception which the Arts and Sciences have met with at this time, from thoſe to whom the heavy burthen of governing human ſocieties is allotted. It is from their care we are to expect the compleat tribunal where all diſputes in this matter can be accurately decided, and all things be rendered truely uſeful ; I mean the inſtitution of a Laboratory ; where the ſight, grinding and poliſhing ; where the air, liquid, and dry diſſolvents and alſo fire in all its degrees, from the electrical to that of the burning-glaſs, may be employed as means to obtain the knowledge of theſe intricate and unknown bodies.

To a ſimilar circumſtance, perhaps, thoſe chemical experiments upon vegetables were owing, which were made many years ago in a certain kingdom ; and though they did not anſwer at that time the intended purpoſe, yet they may at ſome future time be repeated with

advantage,

advantage, when more knowledge in that matter is obtained : But thus much we certainly know by experience, that the mineral kingdom is extremely well adapted to be examined by thefe means. The experiments made by the ingenious Mr. Homberg, with Tfchirnhaufen's burning-glafs, may certainly be carried yet farther, whereby fome doubts may likewife be removed, which ftill remain regarding fome of the effects of his experiments. Thus, we fhould be employed in obferving the phœnomena and drawing conclufions from them, inftead of only fearching for the principles of thofe effects, as naturalifts were formerly obliged to do.

How fatisfied would every lover of fyftems be, if by this means he could get materials properly prepared to compofe a better work, in which he could introduce the few valuable things which are to be found among the old ruins, and leave out all the vague expreffions, together with the diftinctions, that are of no confequence.

When I had, for the above-mentioned purpofe, collected my own obfervations, and thofe of others, I heard of two new books on the fame fubject; they were Mr. D'Argenville's Oryctology, and Mr. Jufti's Mineralogy; for which reafon I laid my manufcript afide, until I had, by the perufal of thofe two works, convinced myfelf that thofe gentlemen had not prevented me from purfuing my plan; for, the former has, in my opinion, endeavoured to

bring

bring us back to a tafte that was formerly in vogue; and which, though we do not defpife, *yet we neglect.* The fecond feems to have hurried himfelf too much, mixing together fome irrefiftable truths, with a greater number of opinions, not yet demonftrated, or mere conjectures; *which is running on fafter with a theory than experiments will permit;* whereby nature, which is the *chief point,* will at the end be loft.

Therefore, that no fondnefs for novelties, in confequence of thefe new works, or others of the like nature, which may hereafter be publifhed, may again divert our attention from the *only method of obtaining any knowledge of the Mineral Kingdom,* which has with fo much pains at length been difcovered, and has already been a little entered upon; I have, prompted either by felf-love, or a more generous motive, publifhed this Effay, even before I have had time and leifure to reduce it into a perfect fyftem: I do not pretend that it is a compleat one, by which we can with certainty divide mineral fubftances, and afterwards reduce them into order. I have chiefly intended it as a bar or oppofition to thofe, who imagine it to be an eafy matter to invent a method in this fcience, and who, *entirely taken up with the furface of things,* think that the *Mineral Kingdom may with the fame facility be reduced into claffes, genera, and fpecies,* as *animals* and *vegetables* are; they do not confider that in the two laft kingdoms of nature there are but feldom,

dom, and never more than two different kinds found mixed together in one body ; whereas in the mineral kingdom it is very common, though it will neverthelefs always remain concealed from every one, however penetrating, *who has not employed himfelf in the compounding or de-compounding fuch bodies*, as far as the prefent knowledge of thefe matters will permit.

So much may be faid in general concerning this Effay ; but now I ought more particularly to inform my readers of the motives why I have now and then deviated from the orders and dif-tinctions hitherto ufed.

Earths and Stones are comprehended in one clafs, becaufe 1. they confift of the fame prin-ciples ; 2. they are by turns converted from one into the other, infomuch that an earth may in length of time become as hard as a ftone, and *vice verfa ;* nor can the true difference be-tween a ftone and an earth be pofitively pointed out by the degrees of hardnefs or foftnefs ; for where is it that the common chalk finifhes, and the lime-ftone begins in the Englifh ftrata ? and how is a clay, whether in water or not, to be diftinguifhed from the foft and unctuous foap-rock, or Smectis.

The divifion of earths into *Vitrefcentes* and *Apyri* is here omitted, fince all of them are, in a due degree of heat, found equally apt, either *per fe*, or by means of fome natural or artificial mixture, to be reduced to glafs equally as well as thofe hitherto called *Vitrefcentes* ; which are nearly the moft refractory in the fire, and ought

to

to be called *Vitrefcentes cum alcali*, if their name was to have any connection with their effects.

Having now fo far overcome the former ignorance, which was the foundation of the knowledge for diftinguifhing the mineral bodies into tranfparent or opaque, hard or foft, we prefer the decifion of the fire, though we ftill labour under the misfortune of not being able to meafure the degrees of fire with fufficient accuracy; for which reafon we always muft fuppofe a *plus ultra* in the experiments by fire.

Sand in reality is nothing elfe than very fmall ftones; therefore, if a feparate clafs were to be made of Sands, another clafs ought to be made, which fhould comprehend Gravel; a third, Loofe Stones; and a fourth, Mountains: This would be a *multiplicatio entium præter neceffitatem*; a fault, which under thefe circumftances may eafily be committed, though not fo foon perceived.

The Saxa muft for the fame reafon be excluded from any fyftem: Otherwife it would be the fame as if a botanift made a difference between the mifletoes, or fuch like vegetables, according to the different genera or fpecies of trees, plants, walls, or rails on which they grew.

Petrefactions, or *Mineralia larvata*, confift of fuch principles as ought to be defcribed in their proper places, without regard to their figure; for which reafon they cannot be enumerated a fecond time. The principal reafon for collecting them, is to acquire a knowledge of fuch bodies of the animal and vegetable kingdoms,

doms, as are not ufually found in their natural
ftate, and in this refpect they belong properly
to the ftudies of the Botanifts, and Zoologifts.
For a Mineralift is fatisfied with a fingle fpecimen of each different fubftance that has taken
the fhape of a vegetable or animal body, and
this only to illuftrate the hiftory of their generation; he leaves it to others to decide if
corals are vegetables, or the habitations of
worms; and thus receives them very uncon-
cernedly, after they have been mouldered to a
chalk, changed into a fpar, or into any other
ftony matter. Neverthelefs, I have in the
Appendix propofed a method for ranging the
Saxa and *Petrefacta* in regard to the œconomical
ufes that may be expected from them.

Slate fignifies or denotes the form alone, and
not its kind or qualities; however, it regards
only its fituation in the rock, and not the texture of its particles; which latter I have always
endeavoured to take notice of, fince fome difference in the effects frequently depends on it.
And, as nothing is great or fmall but by comparifon, it is difficult ftrictly to determine in
what degree of thicknefs or thinnefs a ftone
begins to deferve the name of a flate. Neverthelefs, I would have prevailed on myfelf to
adopt this general name, if the breaking in
thin plates had been the property only of
any particular kind of ftone, but it is by no
means the cafe; becaufe there is found in the
province of Jemteland, in Sweden, a pure
quartz, limeftone, (both folid and fcaly) indurated

rated holes, alum ores, and a great number of faxa, which are all of a plated structure, dividing into leaves as thin as pasteboard : And therefore I make no doubt but all kinds of stones may be found of such a figure in some parts of the world. What confusion would it not occasion if all these different kinds were included under one genus? And are there not solid stones found consisting of the same constituent parts as the slates, which are separately considered in systems?

I could not range the ores according to the different kinds of rock in which they are found ; for instance, the Goose-dung silver ore, the Liver ore, and many more of the same kind, since observations upon the matrices of ores belongs to another branch of the Mineral Science, called *Geographia subterranea*, or *Cosmographia specialis*; in which likewise the clefts, fissures, and veins or loads, that occur in every sort of rock, (in order to promote the circulation and fixation of the mineral vapours) are treated of. In that Science also the *Petræ Parasiticæ* are pointed out, whose number perhaps is not yet known; as likewise the glossy fissures from which the *Mineræ Speculares* have obtained their name. And this is a branch of science, which, in the hands of a skilful mineralist, is likely to furnish us with a perfect idea of the age of every sort of stone, and also of their different gradations between the two accidents, to which all created beings are subject, Composition and Destruction.

Since

Since it has not been ufual to confider the earths and ftones as the fame, in regard to the principles of which they confift, and only different from one another by the greater or leffer hardnefs and coherency of their particles; I hope for excufe, in not being able to profecute this Effay fo far as to point out a particular earth for each kind of ftone. Perhaps alfo fome of the mineral bodies are already fo much hardened, that no earths of this kind are to be found ; or, perhaps, the little knowledge we have of them is owing to the neglect of not collecting earths with the fame earneftnefs as we do ftones and ores. If, therefore, all the earths which are ranged in certain modern fyftems, and there diftinguifhed from one another, in regard to the difference of their colours and places where they are found, had fallen into the hands of Mr. Pott, and been tried by him in the fire, as he has tried the ftones, and been defcribed in his *Lithogeognefia*, it is very probable that we now fhould know thefe bodies better, and their number would certainly be lefs perplexing.

The hints which are here given may, however, tend to promote the intended point for the future ; and then perhaps the earths will be found not to be quite of fo many different kinds as I have here been obliged to divide them into, for want of perfectly knowing their affinities and their feveral origins; becaufe we have ftrong reafons to believe that the calcareous and argillaceous earths are the two principal ones, of which all

b the

the reſt are compounded, although this cannot yet be perfectly proved to a demonſtration.

The *Luſi Naturæ* are not ſeparately treated herein ; they may be found every where in the Eſſay, becauſe the rock cryſtals do not appear to me leſs curious than the indurated marles ; and the kidney ore is often found of a more ſurpriſing figure than the eagle-ſtone. I cannot therefore find any reaſon for forming a claſs of them, nor do I comprehend what others take to be a greater or leſs ſport of nature in the mineral kingdom.

Figured ſtones, or which contain the repreſentations of vegetables, animals, &c. occaſioned by different veins or colours in the ſtones, are in my opinion of ſtill leſs conſequence, and are alſo more difficult to range properly in claſſes ; becauſe people ſeldom, or rather never, agree in their fancies ; but what ſeems to repreſent an object to one, may make a different appearance in the imagination of another, whereby both confuſion and diſputes would ariſe about the bodies of ſuch a claſs. Beſides, a very ſmall advantage, if any, can reſult from ſuch a claſs ; ſince all that the Almighty created has required equally his power, and is alſo equally worthy of our admiration : It gives rather riſe to pedantry, whereby the attention is by degrees diverted from true knowledge to mere trifles, of which both ancient and modern books are ſad inſtances ; and if Mr. Du Fay had not impeded the evil by diſcovering the method of improving the figures upon flints and

and agats, we might still have seen whole collections full of them with such imaginary figures.

Stones that are found in animals and fishes, are partly compounded of phlogiston, salts, and a small quantity of earth, and partly consist of the same matter with animal bones, and can therefore with as little reason have a place in a mineral system as the stones of fruits. Soot, tartar, yeast, and things of such nature, have too great affinity to the vegetable kingdom, and are never to be met with under the surface of the earth; wherefore they may in Botany be considered in the same manner as regules, glasses, and slags are in Mineralogy.

The hair-balls found in animals, and felt, differ from one another in that the former are worked together by means of the peristaltic motion in the bowels of the animals, and the latter by the art of the feltmonger. May not all these stones of animals therefore be ranked among the *relicta animalia* ?

By all this it is very evident, that my chief care has been to treat the mineral kingdom in such a manner, that those whose principal study it is, may avoid every thing unnecessary and superfluous; and by a perfect knowledge of the subjects be brought to consider how to employ them to the best advantage; whereby I hope that the pleasure of collecting minerals will rather encrease than be discouraged. If some objects are thrown out from mineral collections on account they do

not

not belong to them, other collections will be augmented; and thus every thing will be brought into a due order. If some collectors should not immediately be able to comprehend this Syftem or Claffification, it cannot much detriment the fcience; and it is likewife of very little confequence in proportion to the advantage that will accrue to the ftudy by this method, the more it fhall be cultivated and improved.

Ptolomy cannot be fuppofed to have known the value of every book he fent to his great library in Alexandria, and he had doubtlefs no time to clafs them himfelf according to the contents of the different books acquired; however, his love for collecting muft have continued during his life: Moreover, the advantages obtainable from it would have been afterwards difcovered, had it not been difperfed; or, as fometimes happens, had not the collection been fuppofed to gain fome additional value by being made difficult of accefs to the learned.

As foon as we are arrived to fuch a pitch, as by the examination of a mineral body to difcover or know all its conftituent parts, and can affert with certainty that it can be no further decompounded by any method hitherto known; then fuch a body ought, according to the intention of this Effay, to receive its fpecific name, and not before; for otherwife it will be vague and trivial. However, I have not affumed a confidence to do this even with fuch mineral bodies that I in fome refpect can aver

I pretty

I pretty well know, and which have not yet
obtained any fixed name ; as I think it proper
to wait that event, until this Effay has paffed
through that trial which I fincerely wifh it deferves to undergo, equally with others of the
fame nature. Then only will be the proper
time to fix the genera and fpecies, according
to fuch characters as fhall be found the moft
natural.

Meanwhile I flatter myfelf with fo much
fuccefs, that ftudents, who intend to follow this
propofed method, will not be fo eafily miftaken in the fubjects of the mineral kingdom,
as has happened with me and others in following
former fyftems ; and I alfo hope to obtain fome
protectors againft thofe who are fo poffeffed with
the *figuromania,* and *fo addicted to the furface
of things,* that they are fhocked at the boldnefs
of calling a *Marble* a *Limeftone,* and *of placing
the Porphyry amongft the Saxa.*

THE

CONTENTS.

1. Loofe

CONTENTS. xxvii

2. Mineralifed

SECT

c   2. Mineralifed

CONTENTS.

ERRATA.

E R R A T A.

Page 18, line 25, read, *Of pyramidal*, &c.
—— 26, —— 3, for *Drufeæ*, read *Drufæ*.
—— 27, *Antepenultimate line* for *Gypfea*, read *Gypfæ*.
—— 38, —— 32, read *calcareous earth*, &c.
—— 58, —— 31, for *Sect*. I. read *Sect*. L.
—— 67, —— 12, for *Sardea*, read *Sarda*.
——106, —— 21, for *Smelling*, read *Smelting*.
——224,—— 4, read, *Of the others*, &c.
In Sig. T, for p. 173—188, read 273—288.

E S S A Y

TOWARDS A

S Y S T E M

OF THE

MINERAL KINGDOM.

SECTION I.

THE Mineral Kingdom contains all thofe bodies which have been formed under the furface of our earth, whether at the firft creation, or any other time fince that period; and which are ftill daily produced from their original or primary principles, being deftitute of feed, life, or any circulation of fluids *.

* The limits between the three acknowledged kingdoms of Nature are almoft impoffible to be afcertained; whence arifes the difficulty of giving any true definition of them: and indeed it may be queftioned, whether any fuch definition can take place, when we become fo far advanced in knowledge, as to fee clearly the dependence and connexion of all natural bodies into one regular Chain or Syftem.

However, at prefent, it is neceffary that the feveral parts of Natural Hiftory fhould be treated of feparately; and as the

whole

Thefe words, *circulation of fluids,* are to be confidered as a confequence of what is before fuppofed, viz. that the mineral bodies are formed *under the furface of our earth :* and by this particular they are diftinguifhed from the fubjects of the vegetable kingdom; with which, however, they have fome refemblance, fome foffils being ftill produced nearly in the fame manner, at leaft according to our imagination. The clefts, fiffures, and countries *, in the rocks, and in the ftrata of the earth, may be looked upon as anfwering to the tubes in vegetables, and the water as the fluid common to them both. Fire, of whofe nature we ftill know fo little, and which we can hardly determine, whether it belongs to the earth any farther than it exifts, and even muft exift, within a folar fyftem, may, perhaps, be an auxiliary equally neceffary to all the three kingdoms of nature.

On the other hand, we know with a kind of certainty, that if the mineral kingdom is allowed to have exifted before the other two, and to have furnifhed them with materials for their exiftence, it is at the fame time deprived of thofe wonderful and incomprehenfible qualities of life and vegetation, the properties of the animal and vegetable kingdoms.

The defcriptions of minerals cannot be extended farther, in a Treatife of Mineralogy, than to

whole extent of this knowledge can hardly be expected from one man, it may perhaps be executed to more advantage by different perfons : in the mean time, we muft be content with definitions, if they give tolerable ideas of the fubftances defcribed, though they fhould not perfectly coincide with the ftrict rules of logicians. This being premifed, I fhall endeavour in the Text to give fome explanation of my own Effay.

* Countries, an Englifh mine-term for the fides or inclofures of a vein or load of ore, e. g, the country of the load is lime-ftone, killas, &c. &c. D. C.

the

the condition in which they are now found will
permit; for, with regard to fuch bodies as are impoffible to be analyfed or totally decompounded,
no account needs to be given of the principia or
primary atoms, which have concurred to their
formation; fince it may with confidence be afferted, that the Creator made only one kind of
matter, from which every thing we meet with in
this world has been produced.

Minerals have been defined as bodies that grow
by external aggregation of particles; but this definition, on reflection, will hardly be found fufficient
and adequate; becaufe the circles produced by the
annual ftagnation of the fap in the wood of a tree,
and the coats or crufts of a ftalactites, are effects
nearly refembling one another. Should it be faid
that the particles on which the yearly growth of a
tree depends, are not carried by the water, and
merely depofited between the bark and the parts of
the preceding year now become folid, but that they
are juices fecreted from the interior parts thro' the
pores of the tree; I could likewife fhow inftances
in the mineral kingdom, that feem plainly to indicate fome procefs has happened fimilar to the
formation of the callus in a fractured bone; although this happens only at a certain age of the
rock: and who knows whether this obfervation
or procefs does not hold good alfo in the animal
and vegetable kingdoms? But this ftill remains
too great a problem in regard to the mineral
kingdom; and we cannot therefore adopt the exprefs meaning of the " circulation of any juices,"
fince we have another refuge left, that is, *fhrinking*
or *contraction*, and *expreffion* or *fqueezing*, of which
more will be faid hereafter. We take more notice perhaps of what happens in the animal and
vegetable kingdoms, becaufe we are ourfelves

fubject

ſubjeƈt to the ſame alterations, and we likewiſe
outlive the greateſt part of them, therefore we
treat them with more eaſe and conveniency;
whereas the changes which the mineral kingdom
undergoes, are hidden to us, and require many
centuries to compleat them.

I cannot ſuppoſe the mineral bodies to be produced by ſeeds, for want of proof; and I do not
know why the metals ſhould have any preference
in that reſpeƈt. Native or virgin copper and
ſilver are produced in the ſame manner as the
ſtalaƈtites. The water carries along with it the
inviſible particles of lime, copper, or ſilver, and
depoſits them upon other ſubjeƈts, either by
means of an attraƈtive power in theſe, or by ſome
alteration in itſelf, occaſioned by its motion. The
precipitated particles are, at the beginning, very
eaſily ſeparated; but in proceſs of time they cohere very firmly, as is evidently ſeen in the native precipitated or Ziment-copper *, which, after
it has been ſome time taken out of the water, is
partly malleable. The figure which native gold
and ſilver have in their rocks or beds, does not
prove any more than do the metallic iron's or copper's accreting into a moſs-like form in the poor
or rich roaſted ores or reguli at the furnaces; it
gives us rather an idea, how thoſe accidents happen, merely by the ſhrinking and the ſudden cooling of their ſurfaces: and we can then alſo begin
to conceive ſomething of the reaſon why cryſtals
are found in a pebble form, or in looſe nodules, as
the *petrified melons of Mount Carmel*, and the Italian iron ores, in form of balls, &c. without
wanting to have recourſe to the ſuppoſition of a

---

* Called Ziment-copper from its being firſt noticed in a
vitriolic water called ziment, at Herrengrund in Hungary.
D. C.

melting

melting heat, if we only carefully obferve thofe marks, which, befides, plainly fhew their having once been foft, or diffolved.

That mineral bodies are ftill prepared in that large workfhop of Nature, the *Earth*, in the fame manner as thofe which are already full grown feem to indicate, is hardly to be pofitively afferted, fince we yet want fufficient obfervations and experiments thereon.

I will, for inftance, mention the whole Flinty Clafs, of which we have not one obfervation, how they are generated. For if any one pretends to have hit upon the quartz cryftals * in the very inftant of their fhooting, it might be afked, Whether he only depended upon the figure, or if he made fuch experiments thereon as might convince us, that no calcareous earth, either pure or difguifed, was alfo at the fame time mixed therewith? To enumerate the many different ways of generation, which we have any reafon to fufpect, does not properly belong to this work ; befides, it would carry me too far from my fubject, and might alfo furpafs my capacity to explain. I will, however, by mentioning the following opinions, try to fpirit up perfons of more experience and leifure, willing to purfue thefe enquiries.

Precipitation from or by water, is already mentioned, as well as a hint given about the formation of flints. This laft does not fuppofe fuch a foftnefs as that of clay when mollified with water, but a fliminefs, a gelatinous, or a mucilaginous matter, and confequently a more radical folution,

---

* The author ufes the term Quartz criftals for the true criftals; for all foreign authors call figured fpars alfo criftals ; and diftinguifh them by the name of fpath or fpar criftals. D. C.

if

if I may be allowed that expreffion; and this is again to be confidered as a different method.

Another way, and which in our times is much favoured, is that of deftruction *, partly by very violent means, fuch as fubterranean fires, and partly by more mild ones, fuch as the weathering, wafting, or decompounding away; and by this way we have innumerable varieties, and new compofitions. The vitriolic and muriatic *acids* do not lie dormant; and being once let loofe by the faid way of weathering or deftruction, they do not ftop till they are faturated with fomething. Where thefe acids cannot penetrate by themfelves, they are forwarded by the water, which, according to the laws of nature, is almoft in a conftant motion : but the effects of thefe falts ought again carefully to be diftinguifhed from thofe of the water alone, becaufe this latter acts both as a menftruum, for inftance, upon the calcareous earth; and at the fame time by its *vis inertiæ*, heavinefs and motion, wears off or abrades fome particles from folid bodies, carries them along, and depofits them in other places, where thefe particles often acquire a different pofition from what they had before. Are the *Bog-ores* produced of decayed mundics, although no vitriolic matter is found in the waters or tracts around them ? or, Are they to be reckoned a fediment of martial earth diffolved in water alone ? Would it be amifs to fuppofe, that a vegetable mould may of itfelf be changed into iron, fince it is found to yield from a grain to about half of its weight of the faid metal, as the experiments

---

* The original has, " Of deftruction that acts privative," which I have omitted, as it is quite unintelligible to me. D. C.

demon-

demonftrate made upon the turf-moors which
are at the foot of the hills or high rocks in the provinces of Dalarne, Jemtland, and Herjeadalen,
in Sweden? or, that certain vapours have predominated, and ftill fhew their power within certain diftricts, where they attack, and fix themfelves to fuch matters as are moft agreeable to
them; fo that trees, which have been buried in
the earth, in fome of its fubverfions, have met
with martial vapours in fome parts of Bohemia;
the flint-producing principle of Loughneagh in
Ireland, and at Adrianople; with the inflammable fubftance in the ftrata of the coal-pits in England, and at Boferup in the province of Skone in
Sweden? or, that fhells muft, without the deftruction of their calcareous principles, be changed into a calcareous fpar, as at the Balfberg in Skone;
be filled with flint, as at Vernon in France, and
in Siberia; or be penetrated with marcafite, as at
Andrarum in Skone; or with a green copper
ore, as at Jarlfberg in Norway? Silver feems to
predominate at Kongfberg in Norway, as the copper does in the north part of that kingdom: and
the fame kind of ftone in which gold abounds
at one place of the world, may contain none at
all at another place; and other fuch examples occur.

I now leave this important fubject, that I may
not difcover my farther want of knowledge therein; but I take the liberty, at the fame time, to
give this advice to beginners, *viz.* not to conclude, that one mineral body is produced by another, only becaufe they are fituated near one another, if it cannot at the fame time be demonftrated by the analyfis, or by artificial mutation; nor
to content himfelf with making his obfervations
only on collections of minerals, or on heaps of

ores,

ores, but to profecute them to the very workfhop of Nature herfelf, where they may be made with more certainty, though in a narrow compafs; I mean, in the very mines, quarries, and diggings, of the ftrata of the earth. For I myfelf have been aftonifhed to hear that the flint is faid to be produced by a calcareous fubftance, only becaufe it is found in the ftrata of chalk in England, and in the limeftone at Kinnekulle in the province of Weftergottland in Sweden : and I am farther of opinion, that a ftone, feparated from its bed, and thrown loofe on the furface of the earth, does not difcover more marks of increafe, than do bones difperfed in a churchyard.

SECT. II.

The bodies belonging to the mineral kingdom are divided into four different claffes : viz.

1. EARTH, or thofe fubftances which are not ductile, are moftly indiffoluble in water or oil, and preferve their conftitution in a ftrong heat *.

2. INFLAMMABLES, which can be diffolved in oils, but not in water, and are inflammable.

3. SALTS : thefe diffolve in water, and give it a tafte ; and when the quantity of water required to keep them in diffolution is eva-porated, they concrete again into folid and angular bodies.

4. METALS are the heavieft of all bodies hi-therto known ; fome of which are malleable,

---

* The Author, by Earths, does not mean (ftrictly fpeak-ing) only Earths, but includes all the kinds of ftones or fof-fils not inflammable, faline, or metallic. D. C.

and

and fome can be decompounded; neverthelefs, in a melting heat they can again be recovered, or brought to their former ftate,
by adding to them the phlogifton they loft
during their decompofition *.

SECT. III.

The FIRST CLASS.

EARTHS, *Terræ*, are thofe mineral bodies,
not ductile, for the moft part not diffoluble in
water or oils, and that preferve their conftitution
in a ftrong heat.

SECT. IV.

Thefe earths (Sect. III.) are here arranged according to their conftituent parts, as far as hitherto difcovered, and are divided into nine orders.

* Here occurs the fame difficulty in regard to the definitions, as was before (Sect. I.) obferved, becaufe thefe enumerated claffes are likewife blended one with another; and therefore fome exceptions muft be allowed in every one of them:
for inftance, in the firft clafs, the calcareous earth is in fome
meafure diffoluble in water, and pipe clay with fome others
diminifh fomewhat in their bulk, when kept for a long time
in a calcining heat. In the third clafs, the calx of arfenic
has nearly the fame properties as falts; and there is no poffible definition of falt, that can exclude the arfenic, though,
at the fame time, it is impoffible to arrange it elfewhere than
among the femi-metals. In the fourth clafs it is to be obferved, that the metals and femi-metals, perfect or imperfect,
have not the fame qualities common to them all; becaufe
fome of them may be calcined, or deprived of their phlogifton, in the fame degree of fire, in which others are not in
the leaft changed, unlefs particular artifices or proceffes are
made ufe of: fome of them alfo may be made malleable,
when others are by no means to be rendered fo. That the
convex

The First Order.

The Calcareous Kinds, *Terræ Calcareæ*. Thefe, when pure, and free from heterogeneous matters, have the following qualities common to them all :

1. That they become friable, when burnt in the fire, and afterwards fall into a white powder.
2. That their falling into powder is promoted, if, after being burnt, they are thrown into water, whereby a ftrong heat arifes, and a partial folution.
3. They cannot be melted by themfelves, or *per fe*, into glafs in the ftrongeft fire.
4. When burnt, they augment the caufticity of the lixivium of potafhes.
5. They are diffolved in acids with effervefcence, in the following manner ;

   a. The acid of vitriol partly unites with them, and forms a precipitate, which is a gypfeous earth, and partly fhoots into felenitical cryftals with that which is kept diffolved, after a due evaporation,

convex furface metals take after being melted, is a quality not particularly belonging to them, becaufe every thing that is perfectly fluid in the fire, and has no attraction to the veffel in which it is kept, or to any added matter, takes the fame figure ; as we find the borax, *fal fufibile microcofmicum*, and others do, when melted upon a piece of charcoal : therefore, with regard to all that has been faid, it is hardly worth while to invent fuch definitions as fhall include feveral fpecies at once ; we ought rather to be content with perfectly knowing them feparately : however, as this is to be an Effay towards forming a Syftem, I have endeavoured, in moft parts, to follow the ufual rules.

b. With

b. With the acid of common falt they make
a fal ammoniacum fixum, which alfo
partly precipitates itfelf.

c. The acid of nitre diffolves them perfectly, and does not part with them again,
unlefs fome alcaline falt is added.

6. They melt eafily with borax into a glafs,
which fuffers impreffions in a degree of heat
below ignition.

7. They likewife fufe into a glafs with *fal fu-*
fibile microcofmicum with an effervefcence *.

8. They melt the readieft of all kinds of ftones,
with the calces, into a corrofive glafs or
flag.

9. They have alfo fome power of reducing certain metallic earths or calces; for inftance,
thofe of lead and of bifmuth, and likewife,
though in a lefs degree, thofe of copper and
of iron: thus

10. Do they, in this laft mentioned article
(9.), as well as in other circumftances, refemble a fixed alcaline falt; from whence
alfo this whole kind is very often, and properly, called alcaline earths.

11. This whole order of earths is common to
all the three kingdoms of nature; becaufe it
is found in the bones and fhells of animals,
as well as in the afhes of burnt vegetables;
it muft, confequently, have exifted before
any living or vegetable fubftance; and is, no
doubt, proportionable to its univerfal ufe,
diftributed throughout the whole globe.

---

* It is to be underftood, that this effervefcence is alfo made
with the borax, as well as with this *fal fufibile microcofmicum*;
and it is alfo to be obferved, that the glaffes made with thefe
falts are quite colourlefs and tranfparent.

SECT.

SECT. V.

The Calcareous Earth is found,

I. Pure,

   1. In form of powder. *Agaricus Mineralis,* or *Lac Lunæ.*

      a. White, is found in moors and at the bottom of lakes, at Reden in the province of Jemtland, at Timmerdala in Weftergottland, and alfo in the provinces of Smoland, Oftergotland, and ifland of Gottland in Sweden *.

      b. Red, is alfo found in Gottland.

      c. Yellow, is found at Timmerdala, in Weftergottland †.

---

* The white mineral agaric, fo called from its finenefs and lightnefs, like to the vegetable agaric, is found in fuch places, i. e. fwamp moors or peats in England and Scotland, as likewife in the fiffures of the freeftone quarries of Oxfordfhire, Northamptonfhire, &c. but the red and yellow forts I never heard of in England. See Hift. Foff. noft. p. 82. N°. vii. D. C.

† This kind of earth feems to be an impalpable powder of mouldered limeftones abraded and collected by the waters, and is therefore common in the neighbourhoods where limeftones are found; and if the ftone is at fome diftance, which is fometimes the cafe, ftill nothing contradictory appears in this opinion of the origin of this fpecies; fince in that cafe it has only been carried farther by the greater rapidity of a ftronger current of water. When this earth is found in the clefts of rocks, it receives more pompous names; fuch as *Gur, Lac Lunæ,* &c. &c. It burns readily into lime, if it is previoufly ftamped, that it may better cohere: it is then, or in its native ftate, ufed for white-wafhing, but eafily rubs off by the leaft touch. At certain places in the province of Smoland in Sweden, there is found in the moors a white earth, which, by its external appearance, refembles the fpecies here defcribed; but it does not fhew any marks of effervefcence with acids, nor does it burn into lime. It were to be wifhed, that thofe who have an opportunity of getting any quantity of this latter earth, would undertake to examine it better.

SECT.

SECT. VI.

II. Friable and Compact, *folida friabilis* *.
Chalk, *Creta.*

a. White, *Creta alba,* is found in England,
France, and in the province of Skone in Sweden, in which laſt place it is only found adherent to flint.   In the two firſt kingdoms there
are large ſtrata of this ſubſtance, in which
flint is imbedded. This ſeems to indicate, that
the looſe flints, or thoſe diſperſed on the ſurface of the earth, have been by ſome cauſes
carried from their native beds; but, as yet,
no one can prove, that chalk and flint are of
the ſame conſtituent parts.

Chalk is, however, a vague name, alſo
applied to other earths; whence we hear of
chalks of various colours : but I do not
know of any which are of a calcareous nature, except this only kind here deſcribed,
and of which there are no other varieties,
otherwiſe than in regard to the looſeneſs of
the texture, or the fineneſs of the particles.

SECT. VII.

III. Indurated, or Hard, *Terra calcarea indu-*
rata. Limeſtone, *Lapis calcareus.*

A. Solid, of no viſible particles, or not granulated, *particulis impalpabilibus.*
This kind varies in regard to hardneſs and colour, for inſtance,

a. White, from Hull in England.

---

* *Solida friabilis* ſeems contradictory and inexplicable;
however, I ſhall ſtrictly adhere to the Author's definitions,
though never ſo faulty, as I only tranſlate the work. D. C.

b. Whitiſh

b. Whitiſh Yellow, is dug at Balſberg iń Skone in Sweden, and in the Venetiań territoriés.

c. Fleſh-coloured, found in looſe maſſes iń the corn-fields iń the province of Upland in Sweden.

d. Reddiſh-brown, found in the iſland of Oeland, the province of Jemtland, at Rett-wick in the province of Dalarne, and at Kimnekulle in the province of Weſter-gottland in Sweden.

e. Grey, at the ſame places.

f. Variegated with many colours, found in Italy, at Blankenburg, and many other places, and is particularly called Marble *.

---

* Though it may diſpleaſe many, yet I muſt own, I can-not find any characters whereby a marble is to be diſtinguiſh-ed from a limeſtone; and I inſiſt upon it, that nothing but the colours and the texture of the particles diſtinguiſh the kinds of limeſtone. But as Nature has eſtabliſhed no rank by colours, and has made every ſolid limeſtone equally capable of a poliſh, before they are ſpoiled by decaying or decompoſ-ing; it is, therefore, out of this ſpecies of ſolid limeſtone, that ſuch as ſtrike the fancy moſt, ought to be chóſen for orna-ments, under the name of Marble.

It belongs to the ſubterranean geographers to examine, if this ſolid limeſtone is ever found otherwiſe than in ſtrata, and without being mixed with any heterogeneous bodies, that likewiſe have been changed into a calcareous ſubſtance. Here, in the northern parts of the world, it is only found in ſuch a manner as ſhews it was formed in ſtrata, by water's taking up and carrying its particles, and afterwards depoſit-ing them in form of a ſediment, juſt as a ſlime or mud (which is the fineſt particles of pounded rocks) gathers to-gether at the ſtamping mills; and as they are thus formed iń the water, there always are heterogeneous parts along with them. Theſe heterogeneous ſubſtances are, however, in too ſmall a quantity, to be capable of having changed the whole maſs into a calcareous ſubſtance (as ſome pretend); not to mention thoſe circumſtances, which, in other reſpects, make ſuch an opinion very improbable.

g. Black,

g. Black, in the province of Jemtland in Sweden, and in Flanders.   See Sect. xxiii. infra.

SECT. VIII.

B. Grained or granulated limeſtone, *Lapis calcareus particulis granulatis.*

1. Coarſe grained and of a looſe texture. This is called *Salt-ſlag* in Swediſh, from its re-ſemblance to lumps of ſalt ; and is found in the ſilver mines at Salberg, in the province of Weſtmanland in Sweden.
   a. Reddiſh yellow.
   b. White. Both theſe varieties are found in the Salberg mines.
   The grained flux ſpar is alſo ſometimes call-ed *Salt-ſlag.*
2. Fine grained.
   a. White, found at Salberg.
   b. Semi-tranſparent, from Solfatam in Italy, in which native brimſtone is found.
3. Very fine grained.   This is the common limeſtone at Salberg.
   a. White and green, from the mine at Sal-berg, called *Storgrufsan.*
   b. White and black, from the mine at Sal-berg, called *Herr Stans Bottn* *.

SECT. IX.

C. Scaly limeſtone, *Lapis calcarcus particulis ſquamoſis ſive ſpateſis.*

---

* This ſpecies has often as beautiful colours as thoſe com-monly called marbles ; but the texture and coherency of its particles will not admit of a good poliſh.

1. With

1. With coarse or large scales.

    a. White, found at Garpenberg, a copper mine in the province of Dalarne in Sweden.

    It is likewise found at Tunaberg, a copper mine in the province of Sodermanland; but with these different qualities, that it loses in a calcining heat forty per cent. of its weight; and, expofed to the air, gets a brownish efflorescence, a sign that it contains some iron, and is a medium between a limestone and the white iron ore called *Stahlsteine*; nor does it excite any effervescence with acids in its crude state.

    b. Reddish yellow, from Finland.

2. With small scales.

    a. White, from the parish of Tuna in Dalarne, in the marble quarries at Kolmorden in the province of Oftergotland, the parish of Lillkyrke in the province of Nerike, and at Rimito and Pargas in Finland.

3. Fine glittering or sparkling.

    a. White, from Carrara in Italy, and Pargas in Finland.

    b. Of many colours. This variety makes out a great number of the foreign marbles *.

---

* This species of limestone takes a good polish, and is therefore used as marble whenever it is found of a fine colour.

It is besides to be remarked, that the grained and scaly limestones (Sect. viii. and ix.) are found either in veins, or form whole mountains, that shew no strata, nor signs of petrifactions.

SECT.

SECT. X.

D. Lime or calcareous fpars, *Spatum calca-reum.*

1. Of a rhomboidal figure.

  a. Tranfparent or diaphanous.

    1. Refracting fpar, *Spatum iflandicum.*
    This reprefents the objects, feen thro' it, double. It is found at Brattforfs, an iron mine in the province of Wermeland, and alfo in Switzerland and Iceland *.

    2. Common fpar, which fhews the object fingle.

      1. White, or colourlefs.

      2. Yellowifh and phofphorefcent, is found at Jonufwando in Torneo Lappmark in the Swedifh Lapland.

  b. Opaque, *Spatum romboidale opacum.*

    1. White, is found in many places, moftly in clefts, and among cryftallifations.

    2. Black, from Winorn at Kongfberg in Norway.

    3. Brownifh yellow, at Salberg.

2. Foliated or plated fpar, *Lamellofum.*

This has no rhomboidal figure, but breaks into thin plates fo placed as to be not unlike fheets of thin paper, laid over each other.

  a. Opaque white, *Spatum lamellofum opacum,* from Winorn at Kongfberg, and Scaragrufvan at Egeren in Norway.

---

* There are vaft quantities of refracting fpar (a variety of the Iflandic) found in the lead mines of Derbyfhire, Wales, and many other parts of England. D. C.

SECT. XI.

E. Cryſtalliſed calcareous ſpars, *Lapis calca-reus cryſtalliſatus.* Spar Druſen *.

It is compoſed of the laſt mentioned ſpar (Sect.x.), that has formed itſelf exteriorly into ſeveral planes or ſides, wherefrom many different figures ariſe, the varieties of which have not yet been fully obſerved, nor can they be exactly deſcribed. The following are therefore mentioned, only as inſtances of the moſt regular and common kinds, viz.

1. Tranſparent, *Spatum druſicum diaphanum.*

   a. Hexagonal truncated, *Cryſtalli ſpatoſi bexagoni truncati.* This is found at the Hartz in Germany, and at Jonuſwando in Lapland.

   b. Pyramidal, *Pyramidales.*

     1. Dog's-teeth, *Pyramidales diſtincti.* Found at Salberg, and in the iron mines at Dannemera in the province of Upland.

     2. Balls of cryſtalliſed ſpar, *Pyramidales concreti* †.

       Theſe are balls which have Druſen, pyramidal, octaedral, ſpars accreted in their hollows or centers: they are found at Rettwin in the province of Dalarne, and other places ‡.

---

* In my Lectures on Foſſils I have adopted this German term of Druſen into our Engliſh language, for a cluſter of regular figured bodies, as a Groupe conveys the idea of a cluſter only, whether regular or of indeterminate figures. D. C.

† The concave figured ſpar balls in the quarries of Somer-ſetſhire, and other counties in England. Such balls of free-ſtone are not unfrequently found. D. C.

The name Spar is very well known, and only uſed to determine a certain figure, viz. when a ſtone breaks into a
                                  rhomboidal

SECT. XII.

F. Stalactitical Spar, *Stalactites Calcareus.* Stalactites, Stone Icicle, or Drop-ftone.

rhomboidal, cubical, or a plated form, with fmooth and polifhed furfaces, it is called fpar; and as it is thus applied to
ftones of different kinds, without any regard to their principles, one ought neceffarily to add fome term to exprefs the
conftituent parts at the fame time as the figure is mentioned;
for inftance, Calcareous Spar, Gyofeous Spar, Flux Spar,
Shorl or Cockle Spar, &c. This term, however, is not applied but only to earths, and fuch ores as are of the fame figure as the Lead Spar, &c.

All cryftallifed fpars, when broken, fhew the fparry figure
in their particles, and the cryftallifation is to be afcribed to
the empty fpace left by the contraction of the fparry principle: fuch holes filled with Drufen of fpars, are in Swedifh
called *Drake*, or *Drufe-hol* †.

The figure of the cryftals varies more in this genus than
in any other, for which no reafon can be affigned; it ought
not to be afcribed to falts, as long as the prefence of any fuch
cannot be proved: but there are ftrong indications to fufpect,
that other fubftances may likewife have received the fame
property to affume an angular furface on certain occafions.
See Mr. Cronftedt's Introductory Speech at the Royal Academy of Sciences at Stockholm.

Befides, the confideration of thofe figures is a thing of
more curiofity than of real ufe, becaufe no miner has yet
been able to make any conclufion relative to the quantity
or quality of the ores, from the difference of the figures of
fpars found along with them; and the grotto makers never
take any notice of the angles or fides, but think it fufficient
for their purpofe, if they make a fine or glittering appearance at a diftance.

It would, neverthelefs, be well if any one would take upon himfelf the trouble to obferve, whether each fpecies of
fpar has not a certain determinate number of figures or fides,
within which it is confined, in its accretions. This has hitherto been impoffible to do, becaufe all fpecies of fpars have
been confounded together, without regard to their different
principles: though, for my part, I do not think it of any
great confequence.

† What the author fays in the above note is of little confequence to
the Englifh ftudent, as the name of fpar is never ufed with fuch latitude
in our language. All fpars of this flakey texture were by our former
writers, as Grew and Woodward, called Talcy Spars; but that term
now is juftly exploded. See my Lectures, D. C.

C 2   This

This is formed from water faturated with lime, which, while running or dropping, depofits by degrees the calcareous earth which it has carried along with it from clefts of rocks, or from out of the earth. It is therefore commonly of a fcaly, though fometimes of a folid and fparry texture. Its external figure depends on the place where it is formed, or the quantity of the matter contained in the water, and other like circumftances.

1. Scaled Stalactites of very fine particles, *Stalactites teftaceus particulis impalpabilibus.*

   a. Of a globular form, *S. teftaceus globulofus.*

      1. White, the pea-ftone from Carlfbad, in Bohemia.

      2. Grey, *Pifoli'hus, Oolithus,* from Gottland in Sweden *.

   b. Hollow, in the form of a cone, *Coniformis perforatus.*

      1. White, is found every where in vaults made with mortar, and through which water has had an opportunity to penetrate; and alfo in grottos dug in rocks of limeftone.

   c. Of an indetermined figure, *Figura incerta. Sinter.*

      From the cavern called the Baumanshole in the Hartz, the aqueduct at Adrianople, in Italy, and elfewhere.

   d. Of coherent hollow cones, *Conis concretis excavatis.*

---

* Alfo the Hammites, from its refemblance to the roes or fpawn of fifh. It has been exhibited by Authors as petrified roes. The Ketton free-ftone, of Rutlandfhire, is a remarkable ftone of this fort. D. C.

Of

Of this kind is a ftalactitical cruft, which
has formed a ftratum, or rather filled a
fiffure between the ftrata of the earth at
Helfingborg in the province of Skone; it
is of a very fingular figure, refembling
conical caps of paper placed and fixed one
in the other, diminifhing by degrees both
in height and the other dimenfions.

2. Solid Stalactites of a fparry texture, *Sta-*
lactites folidus particulis fpatofis.
   a. Hollow, and in form of a cone, *Coni-*
 formis.
     1. White, and femitranfparent, from
 Chaceline near Rouen in France *.

* In making lime-water (*aqua calcis viva*) one may obferve
how the lime gathers, firft like a pellicle on the furface of
the water, and afterwards, when this breaks, falls down to
the bottom in form of a fcaly fediment, which is called *cre-*
mor calcis : after that, a new pellicle is formed, which likewife falls down ; and in this manner it continues for a long
while, although the lime-water had before been paffed thro'
a filtre. This we may alfo imagine to be the way in which the
works of Nature are performed : whence the ftalactites
commonly is of a fcaly texture, or at leaft difcovers fome
tendency towards it. But a ftalactites of a fparry texture,
fuch as above-mentioned from Rouen, may be fuppofed to
be owing to a more copious principle concurring at once :
and in the fame manner the fparry limeftone and its cryftallifations feem likewife to have been produced, fince they, as
far as I know, are only found in clefts, which, when they
have been filled up with a ftony matter, the Swedifh miners
call *Klyfter*, and *Gangar* or *Veins.* In regard to this, the ftalactites, the fparry limeftone, and alfo its cryftallifations,
might all be ranked under the fame title in a fyftematical defcription, as very little different from one another, if it was
not neceffary in defcribing mines, and other works, to give
them their feparate names : becaufe it is certain, that a piece
which is broken from large fpar-cryftals, or from fparry ftalactites, may in a cabinet pafs extremely well for a common
fparry limeftone, without leaving any fufpicion of its former
figure, before it was broke.

C 3   SECT.

SECT. XIII.

B. Satiated or united with the Acid of Vitriol, *Terra calcarea acido vitrioli faturata.* Gyp-fum. Plafter-ftone or Parget.

This is

1. Loofer and more friable than a pure cal-careous earth.

2. Either crude or burnt, it does not excite any effervefcence with acids, or at moft it effervefces but in a very flight degree, and then only in proportion as it wants fome of the vitriolic acid to compleat the fatu-ration.

3. It readily falls into a powder in the fire.

4. If burnt, without being red-hot, its pow-der readily concretes with water into a mafs, which foon hardens ; and then

5. No heat is perceived in the operation.

6. It is nearly as difficult to be melted by itfelf as the limeftone *, and fhews moftly the fame effects, with other bodies, as the limeftone : the acid of vitriol feems, however, to promote its vitrification.

7. When melted in the fire with borax, it puffs and bubbles very much, and for a long while, during the fufion, owing to the nature of both the falts †.

---

* I have found moft of the gypfeous kind, and particularly the fibrous, to melt pretty eafily by themfelves in the fire.

† When a fmall quantity of any gypfum is melted to-gether with borax, the glafs becomes colourlefs and tranfpa-rent ; but I have found fome forts of alabafter and fparry gyp-fa that, when melted in fome quantity with borax, yield a fine yellow tranfparent colour, refembling that of the beft topafes. This phœnomenon might probably happen with every one of the gypfeous kind. But it is to be obferved, that if too much of fuch gypfum is ufed in proportion to the

8. Burnt with a phlogifton, it fmells of fulphur, and may as well by that means, as
by both the alcaline falts, be decompounded; but for this purpofe there ought to
be five or fix times as much weight of falt
as of gypfum.

9. Being thus decompounded, the calx or
earth which is left, fhews commonly fome
marks of iron.

SECT. XIV.

The Gypfeous earth is found

1. Loofe and Friable, *Terra Gypfea pulverulenta.* Gypfeous Earth, properly fo called,
Guhr.
   a. White, is found in Saxony.

SECT. XV.

2. Indurated, *Terra Gypfea indurata.*
   A. Solid, or of no vifible particles, *Solida
   particulis impalpabilibus.* Alabafter, *Ala-
   baftrum.*

   This ftone is very eafy to faw and cut,
   and takes a dull polifh. It is not always
   found fatiated with the acid of vitriol.

   a. White, alabafter.
   1. Clear and tranfparent, from Perfia.
   2. Opaque, from Italy, and Trapano in
   Sicily.
   b. Yellow.

borax, the glafs becomes opaque, juft as it happens with the
pure limeftone. See the following Treatife on the Pocket
Laboratory, Sect. xxviii.

1. Tranf-

1. Tranſparent, from the Eaſtern countries.
2. Opaque.

SECT. XVI.

B. Gypſum of a ſcaled or granulated ſtructure,
Gypſum particulis micaceis. This is the common Plaſter-ſtone.
1. With coarſe ſcales.
   a. White, is found in the copper-mines
   of Ardal in Norway, where this ſtone
   is the country for the copper-ores.
2. With ſmall ſcales.
   a. Yellowiſh, from Montmartre near Paris.
   b. Greyiſh, from Speremberg in the Mark
   in Germany.

SECT. XVII.

C. Fibrous Gypſum, or Plaſter-ſtone, improperly (though commonly) called Engliſh
Talc by our druggiſts, *Gypſum fibroſum,*
Alabaſtrites.
1. With the fibres coarſe.
   a. White, from Livonia.
2. With fine fibres.
   a. White, is found in very thin ſtrata in
   the alum rock at Andrarum in the province of Skone.

SECT. XVIII.

D. Spar like Gypſum, *Gypſum ſpatoſum.* Selenites. This by ſome is alſo called *Glacies*
Mariæ, and is confounded with the clear and
tranſparent Mica, *Mica alba pellucida* (Sect.
xciv.).

                    1. Pure

1. Pure Selenites.

A, Tranſparent, *Spatum Gypſeum diaphanum.*
   a. Colourleſs, from Swiſſerland.
   b. Yellowiſh, from Mont-martre near Paris.

2. Spar like Gypſum, *Marmor metallicum.*

This ſtone, on account of its heavineſs, which comes near to that of tin or iron, is ſuſpected to contain ſomething metallic; but, as far as is hitherto known, no one has yet been able to extract any metal from it, unleſs ſome traces of iron, which is no more than what all other gypſa contain.

A. Semitranſparent, *Spatum Bononienſe,* the Bononian ſtone or phoſphorus. Its ſpecific gravity is 4,500 : 1000.

B. Opaque.
   a. White,
   b. Reddiſh, are found in Wildeman, at Hartz, and in other German mines.

3. Liverſtone, ſo called by the Swedes and Germans *. See Sect. xxiv.

---

* Mr. Margraff has publiſhed ſome curious experiments in the Memoirs of the Academy at Berlin, about the quality theſe ſpars have to yield a phoſphorus; and has ſhewn, that every gypſeous earth is fit for it, provided metallic particles are not predominant in it: now, as the Bononian ſpar, which is ponderous, is of this ſpecies, and is the moſt fit to be brought to a phoſphorus, it is evident, that no metallic mixture is the cauſe of its weight. Mr. Scheffer, in the Memoirs of the Academy at Stockholm, for the year 1753, has communicated ſome experiments upon a ſtone of this kind from China, which prove that it perfectly agrees with the deſcriptions given in ſeveral books, of a ſtone called *Petuntſe* by the Chineſe, and which, it is ſaid, is uſed in their China manufactories. The phoſphorus of Baldwin illuſtrates Mr. Margraff's experiments. The phoſphoreſcent quality of theſe ſtones is, however, different from that of the ſparry limeſtones

SECT. XIX.

E. Cryſtalliſed Gypſum, *Gypſum cryſtalliſatum.* Gypſeous Druſen, *Druſæ Gypſeæ.*

1. Druſen of cryſtals of pure ſparry gypſum.

   A. Wedge-formed, *Cuneiformes,* are com-poſed of a pure ſpar-like gypſum. See Seĉt. xviii. 1.

      1. Clear and colourleſs, from Switzerland.
      2. Whitiſh yellow, from Montmartre.

   B. Capillary, *Capillares.*

     a. Opaque, whitiſh yellow, from Stoll-berget in Kopparberg Slan in Sweden.

   C. Of ponderous ſpar-like Gypſum, *Marmor metallicum Druſicum.*

      1. Jagged or like cock's combs, *Criſtati·* Theſe reſemble cock's combs, and are found in clefts or fiſſures accreted on the ſurfaces of balls of the ſame ſub-ſtance.

      2. White, from Hartz and Kongſberg in Norway.

      3. Reddiſh, from Wildeman mine in the Hartz.

SECT. XX.

F. Stalaĉtitical Gypſum, *Stalaĉtites Gypſeus. Gips Sinter.*

   This, perhaps, may be found of as many different figures as the calcareous ſtalaĉtites or ſinters. See Seĉt. xii. *c.*

   I have only ſeen the following, viz.

limeſtones and fluors, which is only produced by their being ſlowly heated, and ſeems to ariſe from a phlogiſton, which is deſtroyed in a glowing heat.

1. Of

MINERALOGY.

1. Of no vifible particles, *particulis impalpabilibus*, in French *Grignard*.
A. Of an irregular figure.
　a. Yellow, from the plafter-pits at Montmartre near Paris.
　b. White, from Italy.
This is ufed in feveral works, as alabafter, efpecially when it is found in large pieces; and then it commonly varies in colour between white and yellow, as alfo in tranfparency and opacity.

2. Of a fpar-like texture, *Textura fpatofa*.
A. In form of a cone.
　a. White and yellow, from Trapano in Sicily.
B. Of an irregular figure.
　a. White, from Stollberget in Kopparberg Slan in Sweden *.

SECT. XXI.

C. Calcareous Earth fatiated with the acid of common falt, *Terra calcarea acido falis com-*

---

* What has been before obferved (Sect. xi. xii.) about accreted fpars and finter, may alfo be applied to this fpecies †.

† Gypfeous foffils abound in England. Plafter ftone, granulated and folid, fome fo very fine as to be alabafter, that is, take a furface and politure, are plenty in Derbyfhire and Nottinghamfhire, where are large pits of it, and alfo in moft of the cliffs of the Sevekn, efpecially at the Old Paffage in Somerfetfhire. A very fine femipellucid folid alabafter is found in Derbyfhire. Fibrous talcs, very fine, are found in the fame pits of plafter-ftone above-mentioned, and many other places. Selenites of many kinds abound in England in clays, infomuch that it is needlefs to enumerate the places. Very fine gypfeous Drufen are found in Sheppy-ifle, and the moft beautiful I have ever feen, perfectly pellucid as criftal and large, has been dug from the falt-rocks at Nantwich in Chefhire. The Selenites Rhomboidalis, a rare foffil in other countries, is frequently found in England; but Shotover-hill in Oxfordfhire, is famous for them. I do not know of many of the fpar-like gypfa of Englifh product, but the Ifle of Sheppy affords a kind (to my knowledge peculiar

munis faturata., *Sal Ammoniacum fixum natu-rale.*

This is found,

1. In fea water.
2. In falt pits.

It is formed in great quantities at the bottom of the falt-pans of the falt works. It attracts the moifture of the air *.

SECT. XXII.

D. Calcareous Earth united with the inflammable fubftance, *Terra calcarea phlogifto mixta, feu impregnata.*

Thefe have a very offenfive fmell, at leaft when they are rubbed, and receive their colour from the phlogifton, being dark or black in proportion as it predominates.

SECT. XXIII.

1. Calcareous Earth mixed with phlogifton alone, *Terra calcarea phlogifto fimplici mixta. Lapis fuillus,* Foetid ftone and fpar, or fwine ftone and fpar. Perhaps the fmell

---

* Perhaps fome kinds of limeftones may exift that contain more or lefs of the acid of common falt, though they are not yet difcovered. It is almoft incredible what quantity of fuch diffolved calcareous earth is contained in fea-water ; and from which the teftaceous animals or fhells get the materials for their fhells or coverings. Perhaps Nature has a particular and fecret method of producing a mineral alcali out of the calcareous earth, and has thus laid this earth, as well as the acid of common falt, together in the water, in order to combine them by degrees, and produce the common falt.

culiar and particular to that fmall fpot of ground, and not found any where elfe in the world) fibrous, and always accreting in radiations like a ftar on the Septaria, thence called Stella Septarii. D. C.

of

of this ftone may not be fo difagreeable to every one : it goes foon off in the fire. Its varieties, in regard to the texture, are as follow :

A. Solid, or of no vifible or diftinct particles, *Solidus particulis impalpabilibus.*

    a. Black : the marble dug in Flanders, and in the province of Jemtland in Sweden.

B. Grained, *Particulis granulatis.*

    a. Blackifh brown, from Wretftorp at Skoers in the province of Nerike.

C. Scaly, *Particulis micaceis.*

    1. With coarfe fcales.

      a. Black, at Nas in Jemtland.

    2. With fine glittering or fparkling fcales.

      a. Brown, from Kinnekulle in the province of Weftergottland, and Rettwick in the province of Dalarne.

D. Sparry.

    a. Black,

    b. Light brown,

    c. Whitifh yellow, found in the flate rocks in the province of Weftergottland.

E. Cryftallifed.

    1. In a globular form, dug up at Krafnafelo in Ingermanland *.

SECT. XXIV.

1. Calcareous Earth united with phlogifton and the vitriolic acid, *Terra calcarea phlogifto*

---

* Many of the limeftones of England are of this fection, being extremely fœtid when violently ftruck. In regard to the fœtid fpars, I have had them from the lead mines of Flintfhire in Wales. D. C.

et acido vitrioli mixta. Leberſtein of the Germans and Swedes. *Lapis hepaticus.*

This ſtone ſometimes readily, at other times only when rubbed, ſmells like the *hepar ſulphuris*, or gun-powder. It excites no efferveſcence with acids, and is a medium between the *gypſum* and the fœtid ſtones of the laſt ſection, to which it has, however, generally been claſſed, although no lime can be made from it; whereas they are the fitteſt of all the different limeſtones to be burnt into lime. It is found

A. Scaly.

 1. With coarſe ſcales.

  a. Whitiſh yellow, from Stollen at the mine called Gotteſhulffe in der Noth at Kongſberg in Norway.

 2. With fine glittering or ſparkling ſcales.

  a. Black, is found in form of kernels or balls in the alum-ſlate at Andrarum in the province of Skone *.

---

* The method that Nature takes in combining thoſe matters which compoſe the liver-ſtone, may, perhaps, be the ſame, as when a lime-ſtone is laid in a heap of mundic, while it is roaſting: becauſe there the ſulphur unites itſelf with the lime-ſtone, whereby the limeſtone acquires that ſmell common to liver of ſulphur, inſtead of which the vitriolic acid alone enters into the formation of gypſum. How the ſulphur combines itſelf may likewiſe be obſerved in the ſlate-balls or kernels from Andrarum alum mines, where it ſometimes combines itſelf with a martial earth, of which this ſlate abounds, and forms therewith pyrites within the very ſlate-balls. I have called this ſpecies the Liver-ſtone, notwithſtanding that name, by reaſon of its colour, has before perhaps been given to ſome other kind; but as (in my opinion) the colour is a circumſtance of very little or no importance in mineralogy, ſo as not to deſerve to be taken notice of, in preference to other characters of more conſequence, I hope my boldneſs herein will be excuſed. The fœtid or ſwine-ſtones, and the liver-ſtone, are, in regard to the ſtructure

of

SECT. XXV.

E. Calcareous Earths blended with an argillaceous earth, *Terra calcarea argilla intimè
mixta.* Marle, *Marga.*

1. When crude, it makes an effervefcence
   with acids, but
2. Not after having been burnt ; by which
   operation it is obferved to harden, in proportion as the clay exceeds the calcareous
   fubftance.
3. It eafily melts by itfelf into a glafs, and
   even when it is mixed with the moft refractory clay.
4. It is of great ufe in promoting the
   growth of vegetables, fince the clay tem-
   pers the drying quality of the calcareous
   earth.
5. When burnt in a calcining heat, it readily attracts water: and, expofed to the
   air, in time, it falls into a powder.

   The varieties of this kind, worthy to
be taken notice of, depend on the different quantities of each of their component
parts, and on the quality of the clay. I
fhall, for inftance, fpecify the following
examples.

SECT. XXVI.

1. Loofe and compact, *Marga friabilis.*
   This diffolves in water like common clay.

of their parts, fubject to the fame varieties as the other fpecies
of lime-ftone: and it is to be obferved, that a volatile alcali
is commonly fuppofed to have entered into the compofition of
the fœtid ftones, though it has never yet been difcovered by
any experiment.

a. Reddifh

a, Reddiſh brown; dug up in the iſland of Gottland in Sweden.

b. Pale red; dug up at Upſal in Sweden. This, when burnt, is of a yellowiſh colour, and uſed for making the earthen-ware in the potteries at Rorſtrand near Stockholm.

SECT. XXVII.

2. Semi-indurated, *Marga indurata aëre fateſcens.*

It is nearly as hard as a ſtone when firſt dug up, but moulders in the open air. It is moſtly ſlatty, and is not uncommon in the ſlate rocks of Sweden, where it lies between the thick beds of ſlatty limeſtone, and is is alſo found by itſelf, forming very thick ſtrata. It does not diſſolve in water, till, by a conſiderable length of time, it has mouldered to a powder.

a. Grey.

b. Red.

Theſe are found at Styggforſſen, in the pariſh of Rettwick, in the province of Dalarne.

SECT. XXVIII.

3. Indurated or ſtone marle, *Marga indurata.*

A. In looſe pieces, *Marga indurata amorpha;* by the Germans called *Duckſtein,* or *Tophſtein.*

 a. White, from Woxna in Carelen, and in the river at Nykioping in the province of Sodermanland.

 b. Grey, found in the provinces of Angermanland and Skone.

It

It is formed from a fediment which the water carries along with it.

B. In continued ftrata, *Marga indurata ftratis continuis.* Hard flatty marle.

SECT, XXIX.

F. Calcareous Earth united with a metallic calx, *Terra calcarea metallis intimè mixta.*

Here, as well as in the others, fuch a mixture or combination is to be underftood, as cannot be difcovered by the eye alone, without the help of fome other means.

The fubjects belonging to this divifion lofe the property of raifing an effervefcence with acids, when they are rich in metal, or contain any vitriolic acid. However, there have been found fome that contained twenty or thirty per cent. of metal, and yet have fhewn their calcareous nature by the nitrous acid.

There are no more than three metals hitherto known to be united in this manner with the calcareous earth, viz.

SECT. XXX.

1. Calcareous Earth united with Iron, *Terra calcarea marte intimè mixta.* White fpar-like Iron Ore, *Minera ferri alba.* The *Stahlftein* or *Weifes Eifenerz* of the Germans.

   1. This ore, however, is not always white, but commonly gives a white powder when rubbed.
   2. It becomes black in the open air, as likewife in a calcining heat.

D   3. In

3. In this laft circumftance it lofes thirty
or forty per cent. of its weight, which
by diftillation has been found owing to the
water that evaporates; and it is poffible
that fome fmall quantity of vitriolic acid
may, at the fame time, evaporate with
the water.

4. It is of all the iron ores the moft eafy to
melt, and is very corrofive when melted.
This kind is found,

SECT. XXXI.

A. Loofe, *Minera ferri alba pulverulenta,*
and is the mouldered part of the indurated
fort.

a. Black.

Is like foot. It is found at Wefter-fil-
verberget in Kopparberg Slan in Sweden
among the earth that covers the white
iron ore.

b. Dark brown.

This fomewhat refembles umbre, and
is found to this day at Solfkienfberget in
Norberg in the province of Weftmanland.

SECT. XXXII.

B. Indurated, *Indurata.*

1. Solid, of no diftinct particles, *Solida par-
ticulis·impalpabilibus.*

a. Red, *Minera ferri calcarea rubra.*

Looks like red ochre, or the red hæmatites, but diffolves in the acid of nitre with a great effervefcence. It is
found at Hellefors in the province of
Weft-

Weſtmanland, and at Graſberg in Gran-
gierde in the province of Dalarne.

SECT. XXXIII.

2. Scaly, *Particulis micaceis.*

   a. White, from Naſſau Siegen, and Weſter-
    ſilverberget in the province of Weſtman-
    land.

   b. Blackiſh grey, in Smalkalden, and the
    Old-mines at Helleforſſen in Weſtman-
    land *.

3. Spar-like, *Spatoſa.*

   a. Light brown, from Naſſau Siegen, and
    Smalkalden.

4. Druſen, *Druſica.*

   a. Blackiſh brown, from Smalkalden.

   b. White.

     1. Porous. This is often called *Eiſen-
     blute*, or *flos ferri.* It is found at Weſ-
     ter-ſilverberget.

     2. Cellular, from Weſter-ſilverberget ‡.

SECT. XXXIV.

2. Calcareous Earth united with Copper, *Terra calcarea croco ſeu calce Veneris intrinſecè
mixta.*

---

* In England in the Foreſt of Dean, where it is called
Grey Ore, and at Bigrig Moor in Cumberland.

‡ Theſe kinds, in regard to their texture, are ſo like thoſe
calcareous ſtones we call Limeſtones (Sect. ix.) and Spars
(Sect. x. xi.), that they may be eaſily confounded with one
another, were not the other characters obſerved at the ſame
time.

                                   A. Looſe

A. Loofe and friable, *Pulverulenta five friabi-lis.* Mountain blue, *Cæruleum montanum,* Germanicè, *Bergblau.*

This diffolves in aquafortis with efferves-cence.

SECT. XXXV.

B. Indurated, *indurata.*

1. Pure calcareous earth mixed with calx of copper. Armenian ftone, *Lapis Armenus.*

Such, according to the defcription of authors, ought the nature of the ftone called Lapis Armenus to be, though the druggifts fubftitute in its ftead a pale blue Lapis Lazuli, free from Marcafite.

SECT. XXXVI.

2. Gypfeous Earth united with calx of Copper, *Terra gypfea Venere mixta.* Is of a green colour, and might perhaps be called Turquoife ore, or *Malachites*; though I do not know if all forts of Turquoife ore are of this nature.

    a, Semi-tranfparent, is found at Ardal in Norway *.

---

* By chemiftry we know, that alcaline falts produce a blue colour with copper, which is changed into green, as foon as any acid is added; and from thence the reafon is obvious, why a green colour may be found among calcareous copper ores, viz. when the vitriolic acid is in the neighbourhood of it.

SECT.

3. Calcareous Earth united with the Calx of Lead, *Terra calcarea ceruſſa nativa intimè mixta.*

This is a lead ochre, or a ſpar-like lead ore, which, in its formation, has been mixed with a calcareous earth, and for that reaſon efferveſces with acids.

A. Looſe and friable, *Friabilis.*
   a. White, from Kriſterſberget at Nya Kop-parberget in Weſtmanland.

SECT. XXXVIII.

B. Indurated.
  1. Scaly.
    a. Yellowiſh, from Kriſterſberget †.

SECT. XXXIX.

Obſervations on the Calcareous Earth in general.

The calcareous earth is diſtributed over the whole world in proportion to the great uſe made of it, though it ſometimes is abſorbed and con-cealed in heterogeneous ſubſtances. If it could be proved, that Nature, to perform its works, does not require more than thoſe two active

---

† Both theſe varieties contain a confiderable quantity of lead, viz. forty per cent. more or lefs, and the calcareous earth is as equally and intimately mixed with it, as in the white iron ore (Sect. xxxiii.). Thus may theſe be diſtinguiſhed from other lead-ochres and ſpar-like lead ores, which are much richer in lead, and never efferveſce with acids. Theſe laſt mentioned alſo ſeem to be produced by nature, nearly as the ſpar-like lead ores, and as the *flores Saturni* are formed in calcining a regule of lead.

D 3   agents,

agents, the acid and the alcali, and that the calcareous earth might, under certain circumftances,
be changed into a mineral alcali, as fome have
conjectured; the neceffity of the exiftence of the
calcareous earth is fufficiently obvious. But all
thefe hypothefes I pafs over, fince they for a long
while have been, and will, perhaps, remain for
ever undecided; it being of more confequence to
take notice of the advantages which are to be derived from it in human œconomy, fince it is more
or lefs employed in moft trades. However, I do
not intend to enter into all thofe particulars, but
only to mention how the calcareous earth, when
in its crude ftate, is commonly made ufe of.

When in the form of a loofe earth, (Sect.
v. 1.) it is ufed for white-wafhing; and is
mixed with clay in agriculture: for, according to
Dr. Kullbell's rules of vegetation, its alcaline
quality ferves to unite fat fubftances with water:
befides that it is of a drying nature, and renders
the clay lefs coherent, or, as the farmers fay,
"makes the plowing eafier." Hence this
loofe earth, in fome foreign countries, is called
marle; for, when added to clay, it promotes the
production of marle. The *humus conchacea*, or
fhell or fea-fand, is looked upon as pretty nearly
of the fame quality; but it is unfortunate that
clay for the moft part is fcarce in thofe places,
where the calcareous earth is found in plenty,
and that fometimes more is expected from this
laft than it poffibly can perform. Loofe earth
may eafily be made into lime, if it is previoufly
flacked with water, and made up in moulds.

The indurated calcareous earth or ftones, in
maffes or lumps, are very ufeful in arable land,
becaufe it moulders by degrees on the furface,
and attracting the water, retains it longer than
most

moft other kinds of ftones. The arable lands in the province of Jemtland, at Rettwick in Dalarne, at Kinnekulle in Wefte;gottland, and in other places, which confift merely of a calcareous earth, or a mouldered flate, would fuffer a great deal from the fun, and a dry feafon, if they were not covered with lumps of limeftone.

The art of building cannot be purfued without the ufe of limeftone ; and in this bufinefs alone it is known, and made ufe of under different names.

The folid limeftone (Sect. vii.) commonly found in Sweden, is moftly flatty, and when it is of fuch indifferent colours as not to deferve polifhing, is called in Swedifh *Telgften*, *Alfvarften*, and *Oelandften*, in Englifh fimply Limeftones. The thickeft ftrata are always employed in large works, to which purpofe they are very well adapted ; and the thinner are cut into fquare flabs to pave the floors, and for the ftairs of houfes, and are alfo applied to other ufes. But great care muft be taken in the choice of thefe ftones, fince it will happen that the finer works, made of this kind of ftone, may, in length of time, crack and break into two or more plates, efpecially if they are expofed to the open air, becaufe this ftone is for the moft part fet with fmall partition veins of indurated marle, which moulders in the air. And this is the reafon why the uppermoft ftrata, in thofe quarries, are always rejected, and regarded as a different kind offtone, and are, by the Swedifh workmen particularly, called *Gorften*, in Englifh Rubble-ftone.

When the limeftone is of brighter colours, though fometimes no more than only of a dark brown, it receives the pompous name of Marble : and for fuch works as are to be polifhed, the pieces are always chofen out

of

of the thick and folid ftrata, which lye fo deep under-ground, that they have not been fpoiled by weathering or decaying. This ftone is likewife the moft proper for that purpofe, in preference to all other limeftone, becaufe it is perfectly opaque, and reflects the light from the furface. Moft of the Italian and antique marbles are of this kind, for inftance :

Marmore Nero di Fiandra, —	The black marble of Flanders.
—— *Giallo Antico,* —— —	Yellow, with fome white veins.
—— *di Fiume d' Arno,* ——	Yellow, with black dendrites.
—— *di Fiorenza, Paefino,* }	Yellow, with brown figures, refembling ruins.
—— *di Porto Venere,* ——	Black, with yellow veins.
—— *Nero et Bianco di Car- rara,* —— —}	Black and white.
—— *Tartufato d'Urbino,* }	Pale yellow, with fpots of a blackifh grey colour.
—— *Brocatella di Spagna,*	Yellow white and red.
—— *Palombina antica,* --	Pale yellow.
—— *Alberino di monte Gal- licano,* —}	Olive colour, with deeper coloured crofs-lines and dendrites.
—— *Roffo di fan guifto,* ——	Brownifh red.
—— *Carnagione di Piftoia,*	Flefh-coloured and yellow.
—— *Fior di Perfico di Sara- vezza,* —}	Crimfon, white and grey.
—— *Pavonazzo,* —— }	Reddifh-brown lumps on a whitifh ground.
—— *Bardiglio,* ——	Bluifh-grey.

With infinite more varieties, whofe number is improperly augmented by thofe who for intereftfake collect fpecimens, and likewife by fome virtuofos, who pay too much regard to colours and figures. From the above we find, that the Italian names are for the moft part taken from the colours. When they have a marble from an unknown place, they call it *antico*. Every one that has a number of bright colours, is called
brocatello,

brocatello, or *brocettelato.* The figures are chiefly regarded in the *Paefino di Fiorenza, Alberino di Monte Gallicano,* &c. &c. When fome of the originals are wanted to compleat the whole fet of marbles, they are fubftituted either by others, that have the moft refemblance to them, or by white marbles ftained or coloured; and this is the cafe with the *Marmore di fangue di Dragone.*

To this fpecies of folid limeftone alfo belongs the marble from Blankenburg, which is red, black, and white; likewife that from the province of Jemtland in Sweden, which is black and white, or only black, and the French marbles, viz. *Serfontaine, Antin, Seracolin, St. Baume, Servelat,* &c. &c. which have feveral bright colours.

The fineft folid modern marbles are thofe from Italy, Blankenburg, France, and Flanders. There are alfo marbles dug in Saxony, and other parts of Germany, in Norway, and Sweden; but either they are not of fuch agreeable or ftriking colours, or elfe are of that fpecies which is called the Scaly glittering Limeftone, mentioned Sect. ix.

There are, however, feveral among the abovementioned marbles, which are partly mixed with the fcaly limeftone (Sect. ix.), though not in fuch a quantity as to conftitute the principal part of the ftone, but only as a fubftance which has joined together lumps of the folid limeftone, or elfe filled up its empty crevices or cracks. This kind, however, ought not to be rejected, but might be ufed as a marble, if only fuch pieces were chofen that have the fineft texture : becaufe thofe with coarfe fcales, when polifhed, are of an icy appearancê, as mafons term it, and do not fhew their true colours, by reafon of the femi-
tranf-

tranfparency of their particles, and their different
pofitions in regard to one another, which may
be feen in the marble from the parifh of Perno
in Finland.

Of the fine glittering limeftone (Sect. ix.),
are the following :

Marmore Bianco di Carrara, the white marble,
the *Saligno*, the *Parian*, the white Italian marble;
which, however, is never polifhed when made
ufe of, but only finely ground down ; the *Bigio
antico*, *Porta Santa*, *Carnagione di Verona*, *et di
Siena*, *Tigrato antico*, *Roffo antico*, *Giallo antico
in oro*, *fiorito*, *et Giallo abrufciato* ; every one of
which is fomewhat tranfparent at the edges.

In the parifh of Pargas near Abo in Finland is
found a white marble, which, to judge from the
famples I have feen, gives room to hope that it is
as good as the Italian, when they have got be-
neath the firft ftratum. But the other Swedifh
white limeftones, for inftance, that from Lill-
kyrke and other places, are either of too coarfe
a texture, or fo intermixed with femi-tranfparent
particles, as to give them, at a diftance, a very
difagreeable appearance, as if they were dirty.

The antient ftatuary marble is likewife very
tranfparent : but as this tranfparency is equally
diffufed through all parts of the ftone, it does no
harm, but makes it rather look like alabafter.

While we are on the fubject of marbles, it is
neceffary to obferve, that as the Italians have a
well founded right of giving names to the different varieties of marbles, and of furnifhing us
with famples, both of fuch as are found in their
country, and of foreign ones which in former
ages have been employed there, and now are
called *Antichi* ; it is in regard to thofe collections,
or *Studie*, that moft ftones which take a polifh,
have

have been called Marble, although the Italian mafon himfelf knows extremely well how to diftinguifh a Marble, a Jafper, and a Granite, from one another, giving the two laft names only to marbles of fuch colours as thofe fpecies generally have, when he either cannot get any real ones of thofe harder ftones, or will not give himfelf the trouble to polifh them. This confufion in the names may, however, in regard to this fyftem be tolerated, fince thefe three different fpecies of ftones, viz. the limeftones, the jafpers, and the granites, are here feparately defcribed: but fince they cannot all be worked in a like manner, nor do they equally refift the violence of time, they deferve to be known by the architects in a clearer manner, and by feparate names.

A yet lefs confufion is that of the Saxum, which, tho' compounded of limeftone and ferpentine *, is called Marble, not only when it contains a greater quantity of the limeftone, as the marble from Kolmorden in the province of Oftrogottland, but alfo when the ferpentine predominates, as in the marble called *Pozzevera de Genoua*, and alfo a kind of green marble from Spain, becaufe this kind of ftone is as eafy to cut and work as a true marble, although the ferpentine is fomewhat fofter, and eafier to polifh.

The calcareous fpar (Sect. x.), and its cryftallifations (Sect. xi.), are more difficult to be burnt into lime, than other limeftones; they are therefore of no great ufe in architecture, any further than that they may be employed in making grottos: Nature has alfo made the quantity of this kind proportionable to its ufe.

---

* What our Author calls Serpentine, is a fpecies of nephritic, and of the clafs of Talcs. D. C.

But

But the gypfum or plafter ftone, on the contrary, is of very great confequence in building ; and its ftrata, which are very fparingly diftributed in the earth, are worth fearching for.   If it is true, that the ftrata of the earth are fituated in a regular order throughout the whole globe, as fome affert, and concerning which they have formed fyftems to themfelves, founded upon obfervations made only at fome few places, we might expect to have a confiderable quantity of this ftone ; but there are innumerable experiments yet wanted, before this can be demonftrated.   In the mean while, it may be afked, and with fome reafon, If the gyp-fum is to be fearched for in any other places be-fides thofe ftrata where there is a pofitive proof of their being formed in the middle age, by means of water carrying their particles with it, and depofiting them as a fediment there, and where alfo the vitriolic acid has been prefent ? Likewife, If thefe ftrata ought before to have been fet on fire, whereby the vitriolic acid has been feparated from the inflammable fubftance, and afterwards fixed itfelf to a pure calcareous earth ‡ ?

The miners ufe crude limeftone to make the *hearths* of their iron furnaces, and as fluxes in melting their ores.   The folid and the fcaly lime-ftones are both employed to the former ufe ; but the fcaly (Sect. ix.) is the beft, and next to that the grained limeftone (Sect. viii.).

Thofe who intend to fearch for limeftone to make lime, and are afraid to miftake the white iron ore (Sect. xxx.) for it, ought only to obferve, that the latter always decays in the open air into

---

‡ This whole paragraph, efpecially the latter lines, is very obicure and unintelligible.   D. C.

a black

a black or blackifh brown powder, and becomes alfo of the fame colour in the fire. However, when this iron ore contains only a very fmall quantity of iron, it may be ufed to make lime; though it becomes of a grey colour, juft as when clay is mixed with limeftone, as is the Alfwarften, in which there is always fome mixture.

It feems as if the white iron ore might be ufed with advantage, and preferably to others, in making cement, whofe conftituent parts are always lime and iron; but neither is it apt to concrete, when once mouldered; nor by experiments made for that purpofe, has it difcovered any quality of binding or uniting: we muft, therefore, examine other fubftances, which may better anfwer the intention; and then it will be found, that iron, which is too much in its metallic ftate, is eafily affected by the vitriolic acid, whereby the cement made of it would in length of time be diffolved, and rendered ufelefs; nor, on the contrary, is a perfectly calcined iron of fo much fervice, as when it has fome of its phlogifton left. For inftance, a cement prepared from the flags of a fmith's forge, mixed with lime and coarfe fand, has been found, in fome refpect, to anfwer all the good effects expected, it depending only on time to fhew, if it is durable enough. The *Terra Puzzolana* and *Terras* are nothing elfe than iron ores mixed with a yet unknown earth. Its effect, however, in the cement, may, perhaps, depend only on the iron, which has been reduced into a particular fubftance by means of fubterraneous fires, for their native places retain evident figns thereof.

If the flate in Henneberg, or Kinnekulle in the province of Weftergottland, fhould happen

happen to catch fire, the uppermoft ftratum, which now confifts of a mixture of iron and different kinds of rocks called *Graberg*, in the accounts given of them, they might, perhaps, be changed, partly into flag, and partly into *Terra Puzzolana.*

SECT. XL.

The SECOND ORDER.

The Siliceous Kind, *Siliceæ.*

This filiceous earth is, of all others, the moft difficult to defcribe and to diftinguifh perfectly: however, it may be known by the following characters, which are common to all bodies belonging to this order.

1. In its indurated ftate it is hard, if not in regard to the whole, yet at leaft in regard to each particle of it, in a degree fufficient to ftrike fire with fteel, and to fcratch it, when rubbed againft it, though the fteel be ever fo well tempered.

2. When pure, and free from heterogeneous particles, it does not melt by itfelf, neither in a reverberatory, nor in a blaft furnace.

3. After being burnt, it does not fall to a powder, neither in the open air, nor in water, as the calcareous order does, but becomes only a little loofer and more cracked by the fire, unlefs it has been very flowly, and by degrees, heated.

4. It excites no effervefcence with acids.

5. In the fire it melts eafieft of all to a glafs with the fixt alcaline falt; and hence it has got the name of Vitrefcent, though this name is, properly fpeaking, lefs applicable

to

to this order, than to a great many other earths ‡.

SECT. XLI.

The mineral bodies that are comprehended in this order, are, indeed, somewhat different from one another. This difference, however, on first sight may be discerned; but, in regard to their effects in the fire, and other chemical experiments, it cannot be esteemed of any great consequence, at least while we are no farther advanced in the art of decompounding these hard bodies, and as long as no one has thought it worth the trouble and expence to use those means which are already discovered for this purpose; I mean the burning-glass or concave mirror; and to continue such experiments which Mr.

‡ It is not yet known, if there is any loose earth of this kind to be found, or if the indurated one is produced of a clay, either pure, or mixed with the calcareous earth, which afterwards has been dissolved, in order to produce this: because I have not yet, at least for my own part, found any loose earth that I can suspect to be a siliceous one, except that which remains after stones of this kind are decayed, and which is found in form of a white crust on the surface of those stones that lye to the day, or on the earth. This being afterwards worn off, and carried away by the water, is, perhaps, gathered together in form of strata. In the same manner window-glass likewise moulders in length of time: but it cannot, therefore, be supposed, that any such decayed particles may, without being previously dissolved in some new menstruum, be reduced into their former substance. I am rather inclined to believe, that Tripoli is such a mouldered siliceous earth, and that the method of Nature in producing most of the flinty kind, is such as we do not rightly know, nor have patience to follow, yet imagine that we in some measure imitate it in making of glass, since both these have some effects common with one another.

Pott

Pott has ingeniously begun, as a basis for his *Lithogeognosia.* For want of this, there is no other way left, than to consider these bodies as simple substances (how much soever compounded they may be), in the following manner.

SECT. XLII.

A. Diamond, *Adamas gemma,*
Which,
1. Of all stones, is the hardest.
2. Is commonly clear, or transparent; which quality, however, may, perhaps, only belong to its crystals, but not to the rock itself from which they have their origin.
3. Its specific gravity is nearest 3,500. When brought to Europe in its rough state, it is in form either of round pebbles, with shining surfaces, or of crystals of an octoedral form *.
  a. Colourless, or diaphanous, or the diamond properly so called.
      But it also retains this name when it is tinged somewhat red or yellow. Being rubbed, it discovers some electrical qualities, and attracts the mastic.

* The diamonds commonly crystallize into octoëdral forms, which, however, often are found somewhat irregular, especially when the surface inclines to crystallize, during the shooting of the whole crystal, and also when several of them unite together into a groupe; in which latter circumstance the one hinders the other from assuming its regular form; and of this I have seen several instances. But the octoëdral is not the only regular form which the diamond assumes; I have lately seen a rough diamond, or in its native state, in a regular cube, with its angles truncated or cut off. E.

SECT.

SECT. XLIII.

B. Red, Ruby, *Adamas ruber.* *Rubinus.*
Which, by lapidaries and jewellers, is, in regard to the colour, divided into,

1. The Ruby, of a deep red colour, inclining a little to purple.
2. Spinell, of a dark colour.
3. The Balais, pale red, inclining to violet. This is fuppofed to be the mother of the rubies.
4. The Rubicell, reddifh yellow.

However, all authors do not agree in the characters of thefe ftones.

* Thefe gems are rather too precious to be examined by all poffible experimental means; however they are, by reafon of their hardnefs, and the particular form of their criftals, with more propriety looked upon to be produced from their own feparate principle, being either formed in a fingle drop, or criftallifed out of their matrix, rather than to be ranked among the quartz criftals; for if the heat of the fun, or the climate were the caufe of the hardnefs of the diamonds, why is not a quartz criftal on the coaft of Barbary harder than one from the province of Jemtland in Sweden? and who can affure us here in Europe, if, at the place where the diamonds are dug, there is any kind of rock, or not, which is the bafis or matrix of thefe precious ftones, in the manner as the quartz is of the rock or quartz criftals? The account which Tavernier has given us, about the digging of diamonds at Golconda, agrees with that of the criftals in Jemtland, viz. that they lye bedded in clay within clufters of criftals, and in clefts. Now fuch of our criftals are always the cleareft, as have never been adherent to the rock, and next to them, fuch, as by fome unknown accidents in nature, have been feparated from their bafis; but fuch, as are yet fixed to the rock, are very feldom fit for any ufe: if this, therefore, fhould happen to be the cafe likewife with the diamonds, it is no great wonder, that they do not at the very place take any notice of the rock, and ftill lefs, that they do not bring any of it to Europe. The clufters or groupes, of criftal from Schneckenftein in Saxony, wherein topazes and rock criftals

E are

SECT. XLIV.

B. Saphire, *Saphyrus gemma.*
It is tranfparent, of a blue colour, and
is faid to be in hardnefs next to the ruby,
or diamond.

are found promifcuoufly mixed, having each their different
forms, colours, and hardnefs, furnifhes a proof that nature
forms the fo called precious ftones or gems, each from its particular matter or principle.

The round diamonds may be fuppofed to have undergone
the fame fate with fome of the rock criftals, viz. to have
been, by changes in the earth, broken from their beds, and
by the agitation of waters, ground and rubbed againft one
another, until they have been rounded or reduced to this
form; fince they are moftly found amongft fand, and are difcovered in places worn down by heavy fhowers of rain.

The ruby is criftallifed into an octoëdral form, as well as
the diamond *, and differs alfo very little from it in hardnefs
and weight ; I have, therefore, confidered thefe two, as being of one and the fame kind, and that with as much right
as others, who have ranked them under the rock criftals,
which laft are more regular than any other earthy fubftance,
as they affume, during their criftallifation, a determined
form, viz. the hexagonal, with a point at one or both ends.

A fort of diamond is found, which is faid to be very foft,
and is called the Jargon †, but this fort is ftill unknown to
me, nor have I found that any experiments have been made
relating to its hardnefs and principles.

I have feen in the collection of the mine-mafter, M. Van
Swab, a diaphanous actoëdral criftal of fluor, which, according to thofe, who only mind the figure, ought to be
called a Diamond.

* What I have obferved about the octoëdral form of the diamonds,
may alfo he applied to the rubies. Befides, the rubies are alfo fometimes
found of irregular hexagonal figures. E.

† The Jargon, fo called by the Englifh jewellers. Its natural fhape is
not yet known, it being found in form of pebbles in the Indies, where it
is fplit into thin pieces, and thus fent to Europe. The jargons are of different colours, viz. white, light yellow, and brown. According to
fome lapidaries, they come neareft to the faphires in hardnefs ; and as
they have, when cut and polifhed, a great refemblance to the diamond,
they are alfo by fome called Soft Diamonds, and one may eafily he im-
pofed upon in purchafing thefe for true diamonds, when they are made
up in any fort of jeweller's work. E.

In

In confequence of the ignorance I plead in regard to thefe forts of ftones, I have given this a place by itfelf.

Saphires are faid to be found in Alfatia, at St. Amarin, but accounts of this kind are in general not to be depended upon, as the fluors are frequently met with in collections and the druggifts fhops under the name of faphires, when they are of a deep blue colour; not to mention that the quartz is always termed a precious ftone, whenever it is found clear, and of a fine colour. The faphire is faid to lofe its blue colour in the fire. Thofe which are but a little tinged are called white faphires. The faphire is feldom found of a very deep blue colour, and free from parallel flaws which run through it †.

SECT. XLV.

C. Topaz. *Topazius, gemma.*

This is a precious ftone, which, when rough and perfect, is fold in a criftallifed

---

† The faphires in their rough or native ftate criftallize moft generally like two oblong hexagonal pyramids pointed at their tops, and joined at their bafes : yet they are fometimes found of an hexagonal columnar form. In the fire they lofe their blue colour.

I have found fome of the deep blue faphires, and fome of a milky colour, which, when looked through, varied their colours in the fame manner as the milky or bluifh opals (Sect. iv. 2, 3.) : this is however no reafon why thofe opals fhould be ranked under the name of faphire, and lefs fo, fince there are alfo agats found of the fame quality (Sect. ix. in the note) it might rather give rife to a queftion, whether the name of *milky or bluifh opal* is not to be confidered as a vague term, fince that principal quality is found in ftones of a fomewhat different nature, tho' they all belong to the flinty order ? E.

form

form. At Schneckenftein in Saxony, thefe criftals are found of a prifmatical oct-oëdral form, with no points, but flat, and with fome facets at the top; however without doubt the oriental topazes have another figure *.

Experiments by fire have been made on the Schneckenftein topazes by Mr. Pott, as may be feen in his *Lithogeognefia*.

To this kind I refer

a. The pale yellow topaz.

Which is nearly uncoloured, and is found at Schneckenftein.

b. The yellow topaze, from Schneckenftein.

c. Deep yellow, or gold coloured topaz, or oriental topaz.

d. Orange coloured topaz.

SECT. XLVI.

E. The yellowifh green topaz, or *Chryfolite*.

Is of a grafs green colour, and may perhaps belong to fome other fpecies, which might be difcovered, if it could be obtained rough, or in its matrix, and large

---

* I have got fome rough Brazil topazes, which are prifmatical, and of a rhomboidal quadrangular figure, pointed at one end; but as thefe feem to be broke off at the other end, it is very likely, that they, as well as many other criftals, may be pointed at both ends, when nothing has impeded them during their criftallifation.

Befides thefe, I have got fome fragments of other topazes, likewife faid to be from the Brazils, which are all of them prifmatical, but plainly fhew, that fome are pentagonal, and other regular hexagons with points.

The topazes lofe their colour in the fire, but fome of them turn red in a certain degree of heat, and are therefore very much ufed inftead of the pale rubies, and even are often fold as fuch.

enough

enough or in such quantity as is neceſſary for experiments to be made.

F. The yellowiſh green and cloudy topaz, the *Chryſopraſe.*

This is perhaps the ſubſtance which ſerves as a matrix to the chryſolite: for thoſe I have ſeen of this kind are like the clear-veined, called in Swediſh milk criſtal, and quartz, which is of the firſt degree of criſtalliſation.

SECT. XLVII.

E. Bluiſh green topaz, or the Beryll.

This varies in its colours, and is called, when

1. Of a ſea-green colour, the *aqua marina.*
2. When more green, the Beryll.

They are found in the ſtream-works in Saxony and Bohemia, in form of pebbles, or round pieces.

SECT. XLVIII.

D. Emerald. *Smaragdus gemma.*

Its chief colour is green, and is tranſparent. I believe it to be, or to have been, a criſtal of its own ſeparate principle, ſince in its qualities it differs both from the above-mentioned, and from the rock-criſtals; but I cannot poſitively aſſert this, ſince I know no more of it, than that it is the ſofteſt of precious ſtones, and that, when heated, it is phoſphoreſcent, like the fluors; and what in ſome cabinets is given out for its matrix, and ſaid to come from Egypt, is nothing elſe than a deep green

E 3   cockle

cockle-ſpar *, of which colour we likewiſe find cockle, or ſhirl, in the iſland of Uto, near Stockholm, and at Norbery, in the province of Weſtmanland.

. Mr. Maillet informs us, that in former times the beſt emeralds were found in Egypt †.

SECT. XLIX.

Obſervations on the preceding bodies called, precious-ſtones, or gems.

I have before mentioned the reaſons, why I give theſe their ſeparate places from the following

---

* The original Swediſh has *Skiorl ſpat*, that is cockle or ſhirl ſpar, of which ſee fully Sect. lxxiii. infra. but the German tranſlation terms it, *ein Schoen ſpat*, a fine ſpar; that is to ſay, (in this ſenſe) a fine fluor. The ſchorl, or ſchirl, of the Germans, is a ſubſtance called by our Corniſh miners *Cockle*. I have therefore in my lectures on foſſils adopted the names of cockles or ſhirls for this ſubſtance in the Engliſh language, D. C.

† The emeralds, in their rough or native ſtate, conſiſt of hexagonal columns, moſtly truncated at both ends, though ſome of them now and then may be found facetted at the ends. I have ſamples of both tranſparent graſs-green, and light green, which in a gentle heat become colourleſs, but white and opaque in a ſtrong fire, without the leaſt mark of any fuſion.

When criſtalliſed cockle, or ſhirl, is found of a green colour, tranſparent, and free from cracks or flaws, it is commonly called emerald by the jewellers, though it is generally of a deeper colour than the true emeralds, and alſo wants its luſtre : and hence it is, that the cockle-ſpar from Egypt is called the mother of the emeralds.

However it may be, that this cockle was in antient times faſhionable in Egypt, under the name of emerald, though now-a-days it is not ſo much valued as the emerald of this (the ſiliceous) kind. See Section lxxv. Note under the cockle.

ftones, among which they elfe might have been ranked equally as here, as they are already in other fyftems : to which I venture to add, that, as a naturalift, I cannot conquer that general weaknefs of valuing them fo highly ; for befides their furprifing hardnefs, and fine colours, that pleafe the eye, it is not without foundation, that they might be thought applicable to every ufe for which the filiceous kind is employed, if they were to be had in large quantities : and by this alone it is they deferve to have the preference of the other ftones of this order. In regard to the colours, it is to be obferved, that thofe of the ruby and emerald are faid to remain in the fire, but that the colour of the topaz flies off : whence it is ufual to burn that gem on purpofe that it may be made ufe of inftead of the diamond, as it is harder than the quartz cryftal. The colours of gems are commonly fuppofed to depend on metallic vapours : but may they not more juftly be imagined to arife from a phlogifton, united with a little metallic or fome other earth? becaufe we find that metallic earths, which are perfectly well calcined, give no colour to any glafs, and that the manganefe, on the other hand, gives more colour than can be afcribed to the fmall quantity of metal which is to be extracted from it. (Section cxiii.)

The phlogifton may perhaps have more difficulty to efcape through the pores of the hardeft of the precious ftones, if it is true, that the property of lofing the colour is in proportion to their hardnefs, as fome authors feem to indicate, by affuring us, that none but the coloured diamonds and the rubies keep their colours in the fire ; but in this circumftance I likewife want experience, but hope to fee it illuftrated

by

by thofe who may happen to get an opportunity
of difcovering the true methods to deliver the
world from the many ambiguities and diftinctions, which have been made on this fubject,
and which perhaps are all formed upon as great
reafon, as thofe we ftill ufe in diftinguifhing the
oriental and occidental gems, which fignifies in
other words, no more than hard and clear, or
foft and flawy, deeper or paler, or of good or bad
colours *.

* To the precious ftones belong alfo the jacinths, or
hyacinths, which are criftals harder than quartz criftals,
tranfparent, of a fine reddifh yellow colour, when in their
full luftre, and formed in prifms pointed at both ends :
thefe points are always regular, in regard to the number of
the facets, being four on each point, but the facets feldom
tally : the fides alfo, which form the main body, or column,
are very uncertain, in regard both to their number and fhape,
for they are found of four, five, fix, feven, and fometimes of
eight fides : further, the column or prifm is in fome alfo fo
comprefied, as almoft to refemble the face of a fpherical
facetted garnet. Thefe criftals lofe their colour, become
white, and do not melt in the fire, by which qualities
chiefly they may be diftinguifhed from garnets, (Section
lxviii. 3.) which are likewife fometimes found of a colour not inferior to the true jacinths. The author had not,
at the time when he wrote this Effay, feen the true jacinths,
but mentions in Section lxix. C. c. that the reddifh yellow
garnets from Greenland, are fold by the jewellers for jacinths ;
fo are likewife the Eaft Indian garnets of the fame colour,
and, what is ftill more, there are fome jewellers that do not
know the true diftinctions between a jacinth and a garnet at
all, but buy and fell the garnets for jacinths, when they are
of a fine reddifh yellow colour : this muft in particular be
owing to the fcarcity of the true jacinth.

Mr. Cronftedt has fince informed me by letter, that he had
lately got fome jacinths of a quadrangular figure, which did
not melt in the fire, but only became colourlefs ; this confirms what I have already mentioned about the jacinths I
tried, and which are above defcribed. E.

SECT.

SECT. L.

E. Quartz. *Quartzum**.

This ftone is very common in Europe, and eafier to be known than defcribed. It is diftinguifhed from the other kinds of the filiceous order, by the following qualities.

1. That it is moft generally cracked throughout, even in the rock itfelf, whereby

2. As well as by its own nature, it breaks irregularly, and into fharp fragments.

3. That it cannot eafily be made red hot, without cracking ftill more.

4. It never decays in the air

5. Melted with pot-afhes, it gives a more folid and fixed glafs than any other of the filiceous order.

6. When there has been no interruption in its natural accretion, its fubftance always criftallifes into hexagonal prifms, pointed at one or both ends.

7. It occurs in clefts, fiffures, and fmall veins in rocks. It very feldom forms large veins, and ftill feldomer whole mountains, without being mixed with heterogeneous fubftances.

SECT. LI.

The Quartz is found,

1. Pure, *Quartzum purum.*

A. Solid, of no vifible particles with a glofsy furface. *Particulis impalpabilibus fuperficie polita.* Fat Quartz.

---

* I fhall adopt this name of quartz in Englifh as it has already gained accefs into the other European languages. D. C.

c. un-

a. Uncoloured and clear, *Diaphanum.*

Is found in the copper mines in the north part of Norway, and in Siberia. This has no criftallifed form, but is neverthelefs as clear as quartz criftals of the beft water.

b. White, the common fat quartz.

c. Blue, from the ifland of Uto in the province of Sodermanland.

d. Violet, from the ifland of Uto.

B. Grained, *Textura granulata.*

a. White, from the gold mines at Adelfors in the province of Smoland, and the copper mines at Lovifeberg in Weftmanland.

b. Pale green, from Adelfors.

C. Sparry quartz, *Textura fpatofa.*

This is the fcarceft, and ought not to be confounded with the white Felt-fpat Sect. lxvi. being of a fmoother appearance, and breaking into larger and more irregular planes.

a. Whitifh yellow, from the gold mines in Hungary.

b. White, from the ifland of Uto.

SECT. LII.

D. Criftallifed Quartz, *Quarzum criftallifatum.* Rock criftal. Quartz criftal. *Criftallus montanus.*

Its figure is already (Sect. L) defcribed, and, in regard to the colours, the following varieties occur.

1. Opaque, or femi-tranfparent, *Criftallus opacus vel femi-diaphanus.*

a. White, or of a milk colour.

b Red

b. Red, or of a carnelian colour, from Oran in Barbary.

c. Black, from the fame place.

2. Clear, *Diaphanus.*

　a. Blackifh brown, fmoaky topaz, or *Raunch Topas* of the Germans, is found at Egern in Norway, and at Lovifa in Finland.

　b. Yellow, found in Bohemia, and fold inftead of topazes.

　c. Violet, the amethift, from Saxony, Bohemia, and Dannemora in Upland.

　d. Uncoloured, Rock Criftal, properly fo called, from Bohemia: alfo from the province of Jemtland, and many other places; when thefe coloured criftals are not clear, they are called Flufs, for inftance Topaz-Flufs, Amethift-Flufs, &c. &c.

S E C T. LIII.

2. Impure Quartz, *Quartzum heterogeneis intimè mixtum.*

A. Mixed with iron, in form of a black calx, *Quartzum calce ferri atro intrinfece mixtum.* This is black, of a gloffy texture, and contains a great quantity of iron.

　It is found at Staf's iron mine in Sodermanland, and at Gierdefioftrand in the parifh of Offerdal in Jemtland, at which laft place the iron alfo difcovers itfelf by its ruft in the cracks of the ftone.

B. Mixed with copper in form of a red calx, *Quartzum croco Veneris mixtum.*

Red

Red, and is found in Sunnerſkog's copper mine in the province of Smoland *.

SECT. LIV.

F. The Flint, *Silex pyromachus.* *Lapis corneus,*
or *Hornſtein* of the Germans.

This is equally common with the Quartz,
and it is full as difficult to deſcribe it; eſpecially as it forms a kind of intermediate ſubſtance between Quartz and Jaſper, both

---

* That the colour in theſe latter bodies depends on metals,
is eaſily proved by metallurgical Eſſays, and the reſemblance
they have with compoſitions of glaſs, made on the ſame principle; but the ſame cannot be affeſted of the precedent coloured quartzes (Sect. li. lii.), before it be evidently demon-
ſtrated.

It is very likely, that a quartz may be found which is intimately mixed with a calcareous earth, and ſuch is, perhaps,
the Hungarian ſparry quartz (Sect. li. *C.*), which, however,
I recommend for further examination.

The quartz in general, and eſpecially its criſtals, are very
commonly thought, when yet in their ſoft or diſſolved ſtate,
to have included within them ſome vegetables, for inſtance,
graſs and moſs. This I cannot abſolutely deny, but I muſt
at the ſame time obſerve, that it deſerves to be carefully examined, if that, which is ſhewn as a graſs, be not an aſbeſ-
tos, or a ſtriated cockle, and the moſs, only branched vacui-
ties filled with earth, which, by their being ramoſe, bear a
vegetable appearance : it is very common in agates, and
makes them of leſs value than otherwiſe they would be; this is
moſt generally the caſe with thoſe ſtones, which are ſhewn as
including vegetables, and for my own part I have never been
ſo fortunate as to meet with any others.

When the rock criſtals are ſemi-tranſparent, or intermixed
with opaque veins, they are, by the Swediſh lapidaries, called
Milk criſtals ; when they are found in form of round pebbles, which is occaſioned by their being toſſed about and
rubbed againſt one another by floods, or by the ſea, they
are called by the Engliſh lapidaries, Pebble cryſtals. They
come from the Indies, Siberia, and other places, but theſe
cannot be ranged ſeparately, for evident reaſons, or otherwiſe
for reaſons already mentioned in their proper places.

which

which it fo nearly refembles, that it is not
eafy to point out fuch characters as fhall readily diftinguifh it from them. The beft
way, perhaps, will be to fpeak of its properties comparatively; and then we may fay
that,

1. It is more uniformly folid, and not fo
   much cracked in the mafs as the quartz;
   and,

2. It is more pellucid than the jafper.

3. It bears being expofed to the air, without
   decaying, better than the jafper, but not
   fo well as the quartz.

4. It is better for making of glafs than the
   jafper, but is not quite fo good as quartz
   for that purpofe.

5. When ever there has been an opportunity in this matter of its fhooting into
   criftals, quartz criftals are always found
   in it; juft as if the quartz had made one
   of its conftituent parts, and had on certain circumftances been fqueezed out of
   it; this is to be feen in every hollow
   flint, and its clefts, which are always filled
   up with quartz.

6. It often fhews moft evident marks of
   having been originally in a foft and flimy
   ftate.

   The feveral varieties of this fpecies
have obtained diftinct names, more with
refpect to their colours, than from any
real difference in their fubftance; but
thefe are ftill neceffary to be retained, as
the only names ufed by jewellers and
others, who know how to value them ac-
cordingly.

SECT.

SECT. LV.

1. The Opal. *Opalus Pæderota.*

It is the moſt beautiful of all the flint kind, owing to the changeable appearance of its colours by reflection and refraction, and muſt therefore be deſcribed under both theſe circumſtances.

a. The Opal of Nonnius, the Sangenon of the Indians.

This appears olive-coloured by reflection, and ſeems then to be opaque, but when held againſt the light, is found tranſparent, and of a fine ruby red.

That opal is ſuppoſed to have been of this kind, which Pliny mentions in his Natural Hiſtory, chap. 307. ſect. xxi and which he ſays, was in the ſenator Nonius's poſſeſſion, who rather ſuffered baniſhment, than part with it to Antony.

This ſtone was in Rome at that time valued at 20000 ſeſterces. But the ſtone here particularly deſcribed, was found in the ruins of Alexandria; it is about the ſize of a hazle-nut, and was bought for a trifle of a French druggiſt, named Roboly, and preſented to the French general conſul Lironcourt, who afterwards offered it to ſale in ſeveral places, for the ſum of 40,000 rixdollars. See Haſſelquiſt's Travels to the Eaſt, under the article of Opal. *

* This very ſtone was in the year 1763 in the poſſeſſion of his excellency the duke de Nivernois, then ambaſſador to the Britiſh court, and I have often been honoured by his excellency of having it for ſome days in my poſſeſſion. D. C.

There

There is, however, another of the fame kind in Sweden, which by reflection appears rather brown; but by refraction is red, with violet veins.

b. The white opal. Its ground is white, of a glafs-like complexion, from whence are thrown out green, yellow, and bluifh rays; but it is of a reddifh or rather flame colour, when held againft the light.

1. Of many colours. The oriental opal.

2. Of a milky colour *, from Eibenftock, in Saxony.

3. Bluifh, and femi-tranfparent. This is not fo much valued, as thofe which are

---

* I have lately got a fmall piece of pfeudo-agate, from the Eaft-Indies, which is of a yellowifh brown, and pale blue, or rather milky colour, with a fhining brightnefs, exactly like that of the milky opals of this fection, and received alfo fome other fpecimens found at St. Georgio, near Turin, in Piedmont, there called by the name, *Pfeudo agate* (Baftard agate), a name which feems very well adapted to this ftone, fince in every refpect, hardnefs excepted, it comes neareft to the agates ; becaufe, 1. It is tranfparent in the fame degree as agates, and varied with red and grey colours, interfperfed fometimes with white opaque veins, or rings, and black dendritical figures. 2. It is of a very fine and fhining texture, when broke, rather fuperior to that of the agates, but fo foft, that it fcarce yields any fparks, when ftruck againft fteel; and does not admit of any polifh, but what is inferior to the luftre of its natural texture : however, it flightly marks common window glafs. 3. When broke through the dendrites, it is as fmooth and fhining at that place as in any other ; and thefe dendrites vanifh in the fire, without leaving any pores in the ftone. 4. It does not melt before the flame, by the blow-pipe, but becomes perfectly white and opaque. 5. Nor is it fufible even with borax. 6. It does not ferment with the acid of nitre.

Some-

more opaque, becaufe it is eafier to be
imitated by art †.

S E C T LVI.

The Cat's Eye. *Pfeudopalus.*

This ftone is opaque, and refleȼts green
and yellowifh rays from its furface, and is
found in Siberia *.

Sometimes this ftone is furrounded with a white cruft, like
common flints in the ftrata of chalk; which cruft has likewife
the faine effeȼt as that of the flint, when this laft-mentioned
has been previoufly freed from the adherent chalk; viz. 1. It
does not ferment nor diffolve in the acid of nitre ; 2. is not
fufible by itfelf in the fire ; 3. but melts pretty eafily with
borax, though without any effervefcence, contrary to what we
obferve with ȼalcareous fubftances ; and thus borax will diffolve
a quantity equal to about three quarters of its own bulk,
thg h not without difficulty, efpecially towards the end of
the operation ; but the glafs becomes quite clear and colourlefs, inftead of growing white and opaque, as with calcareous
fubftances. E.

† Not only this, but alfo fome of the other kinds of opals,
have been well imitated by art, there being found compofitions of glafs, which fhew very different colours by refraction from what appear by refleȼtion. A curious antient one
of this kind is to be feen in the Royal Abbey of St. Dennis,
near Paris, which is green on the outfide, and fhews a fine
ruby colour when viewed againft the light. And lately an
ingenious gentleman at London made fome paftes, which are
of a yellowifh dark brown by refleȼtion ; but fome of which,
when held againft the light, appear of a fine blue colour, and
others either purple, or like hyacints, garnets and rubies.

* The earlier writers on ftones mention other varieties of
this kind ; for inftance, the *Oculus Mundi* §, which, after hav-

§ There are in the Britifh Mufæum at London, three of thefe ftones
called *Oculus Mundi*. The largeft of them is about the bignefs of a
cherry-ftone, though in an oval form. It is opaque, and its colour like
that of a common yellow pea ; it may be fcratched, though not without difficulty, by a knife ; it feems however to leave a mark on common glafs, and does not ferment with the acid of nitre.

When it has lain in water fome hours, it becomes tranfparent, and of
a yellow amber colour. This change begins foon after the immerfion,
and at one end, in form of a little fpot (but in a fmall one of the fame
kind

SECT. LVII.

3: The Onyx. *Onyx Camehuja.* Memphites. This ftone is the hardeft of the flinty tribe, and confifts of differently coloured veins, which run parallel to one another, fometimes in ftraight, fometimes in curved lines. It is found of two forts.

a. Nail-coloured onyx, having pale flefh-coloured, and white lines. From the river Tomm in Siberia.

b. With black and white lines. The Oriental onyx.

The old Romans were accuftomed to cut figures on the ftraight-lined onyxes in relief, which they called *Camehuja*; thefe are ftill counterfeited, and called Camayeu. Thofe which confift of concentric circles were called Memphites; and we have now of this kind cut to be fet in

ing been laid in water, fhines like a piece of red-hot charcoal; the Afteria, which is faid to fhew luminous ftars on its furface, &c. But thefe are no longer to be met with, fince fafhion has given preference to the more tranfparent hard ftones; and it is alfo very difficult rightly to underftand the defcriptions of antient authors, in regard to colours, and their different mixtures with one another ‡.

kind the beginning is round the edges) which increafes by flow degrees until the whole ftone is become uniformly clear throughout: when taken out of the water, it lofes its tranfparency, firft at one end, then gradually over the remainder, until the whole ftone has recovered its former opacity; and this change happens in lefs time than that of its becoming tranfparent. No other experiments have yet been made upon this ftone, becaufe it is fo very feldom to be met with; and thefe are not fufficient to determine exactly of what kind it is. E.

‡ Amongft thefe cannot be reckoned the Tourmaline, fo much renowned of late for its electrical qualities. It is but a cryftallifed cockle, of a green or brown colour, more or lefs deep, fo as to turn to a black, and fometimes to a bluifh colour, when looked through; others appear quite black. [See Sect. LXXV.) E.

The Brafilian emerald feems to have the fame properties of becoming electrified pofitively in one fide, and negatively in the other: it belongs alfo to the Shirl kind. [See Sect. XLVIII. Note †.]

F   rings,

rings, under the name of *Occhi di Gatti*,
which, however, ought not to be confounded with the pfeudopal, (Sect. lvi.)
or cat's-eye.

SECT. LVIII.

4. The Chalcedony, or white agate.

I a flint of a white colour, like milk diluted with water, more or lefs opaque: it
has veins, circles, and round fpots. It is faid
to be fofter than the onyx, but much harder
than thofe agates which are fometimes found
of the fame colour.

a. The white opaque Chalcedony, or *Cacho-
long*, from the Buckarifh Calmucks. This
was firft made known by one Renez, a
Swedifh officer, who for feveral years had
been in that country. The inhabitants
find this flint on the banks of their rivers,
and work idols and domeftic veffels out
of it.

b. Of white and femi-tranfparent ftrata, from
Ceylon.

c. Bluifh grey, from Ceylon and Siberia.

SECT. LIX.

5. The Carnelian, *Carniolus.*

Is of a brownifh red colour, and often entirely brown. Its name is originally derived
from its refemblance to flefh, or to water
mixed with blood.

a. Red, from the Eaft, and Turky.

b. Yellowifh brown, looks like yellow amber, from the river Tomm, in Siberia.
It is faid not to be fo hard as the Chalcedony.

SECT.

SECT. LX.

6. The Sardonyx.

Is a mixture of the chalcedony and carnelian, sometimes stratum-wise, and sometimes confusedly blended and mixed together.

a. Striped with white and red strata : this serves as well cut in cameo as the onyx.

b. White, with red dendritical figures.

This very much resembles that agate which is called the Mocha stone, but with this difference, that the figures are of a red colour in this, instead of black, as in that agate.

I have unwillingly distinguished the onyx, carnelian, chalcedony, sardonyx, and agate, as separate species, since there is no real difference between them, except some inexplicable degrees of hardness; but I have been induced to continue these names for the reasons before given in Sect. liv.

SECT. LXI.

7. The Agate, *Achates.*

This name is given to flints that are variegated with different colours, promiscuously blended together; and they are esteemed in proportion to their mixture of colours, their beauty and elegance. Hence also they have obtained variety of names, mostly Greek, as if the business of the lapidary in cutting of them, and admiring their several beauties and figures, had been derived from that nation alone.

F 2   As

As it ever was and muft be very difficult
to give intelligent defcriptions of colours,
fo we are quite at a lofs to underftand the
meaning of the antients in this refpect; but
this indeed is of little confequence, as we
feem to have the fame right, under the fame
circumftances, of inventing new names for
them; and that in whatever language we
pleafe. Neverthelefs I have defcribed fome
few varieties of thofe which at this time are
the moft common, to ferve as inftances.

a. Brown opaque agate, with black veins,
and dendritical figures, the Egyptian
pebble.

b. Of a chalcedony colour, *Achates chalce-
donifans.*

c. Semi-tranfparent, with lines of a blackifh
brown colour, and dendritical figures, the
Mocha ftone.

This is much efteemed, and makes a
valuable part of fome collections, where
it has a place chiefly for the fake of its
figures, refembling vegetables, animals,
&c. which however are often improved by
art.

d. Semi-tranfparent with red dots, *Gemma
divi Stephani.*

When the points are very minute, fo
as to give the ftone a red appearance, it is
by fome called *Sardęa.*

e. Semi-tranfparent, with clouds of an orange
colour.

f. Deep red or violet, and femi-tranfparent.

g. Of many colours, or variegated.

h. Black.

There

There is in Europe great quantities of moſt varieties of agates *, particularly at Oberſtein in the Palatinate, where they are cut and poliſhed : but they are like-wiſe found in every part of the world. In Sweden there is not yet, as far as I know, more than one ſpecies of agate found; namely, at Gaſebeck, in the pro-vince of Skone, which is of a white and deep red colour.

S E C T. LXII.

8. Common Flint, *Silex communis Pyroma-chus.*

Is really of the ſame ſubſtance as the agate ; but as the colours are not ſo ſtriking

---

* I have lately got a ſpecimen of a hollow agate ball, with pale améthiſts in the inſide, between which is cryſtalliſed a calcareous ſubſtance into a fibrous form. Theſe fibres are parallel, white, ſhining, and very minute, exactly reſembling the fineſt aſbeſt, for which it alſo might be miſtaken, if it was to be judged only by the eye. But by experiment it is found neither to be an aſbeſtus nor a gypſum, which ſometimes ſhoot alſo into a fibrous form, but entirely a pure calcareous ſubſtance. The whole maſs does not adhere together, but is nearly divided into ſmall triangles, which are placed upon one another, ſo as almoſt to form a large figure of the ſame kind. Theſe fibres however, although very minute, may be found by means of a proper manifying-glaſs, to be of an angular figure, like thoſe mentioned in the note at page 82. The ſhape of balls and irregular nodules, is the moſt general form in which agates and flints are commonly found. Neverthe-leſs, beſides what I have ſeen in ſeveral collections in London and abroad, I have likewiſe ſome ſpecimens of native ſilver, from Potoſi, in the Spaniſh Weſt Indies, which run in a grey and blue tranſparent agate, with white opaque veins ; which ſeems to confirm the opinion, that agates may form veins in the rocks, as well as other ſorts of ſtones. [See the note of Sect. LXIII.] E.

or agreeable, it is commonly confidered as a different fubftance.

a. Blackifh grey, from the province of Skone.

b. Yellow femi-tranfparent, from France.

c. Whitifh grey.

d. Yellowifh brown.

When the flints are fmall, they are in England called pebbles; and the Swedifh failors, who take them as ballaft, call them *fingel.*

S E C T. LXIII.

9. Chert, *Petrofilex, Lapis Corneus.* The *Hornftein* of the Germans.

Is of a coarfer texture than the preceding, and alfo lefs hard, which makes it confequently not fo capable of a polifh. It is femi-tranfparent at the edges, or where it is broke into very thin pieces.

a. Chert of a flefh colour, from Carl-Schakt, at the filver-mine at Salberg, in the province of Weftmanland.

b. Whitifh yellow, from Salberg.

c. White, from Kriftiersberg, at Nya Kopparberget, in Weftmanland.

d. Greenifh, from Preftgrufvan, at Hellefors in Weftmanland *.

* There are not yet any certain characters known, by which the Cherts and Jafpers may be diftinguifhed from each other : by fight, however, they can eafily be difcerned, viz. the former, or chert, appearing femi-tranfparent, and of a fine fparkling texture, on being broken; whereas the jafper is grained, dull, and opaque, and has exactly the appearance of a dried clay : the chert is alfo found forming larger or fmaller veins, or in nodules like kernels in the rock; whereas the jafper, on the contrary, fometimes conftitutes the chief fubftance of the higheft and moft extended chains

of

of mountains. The chert is likewise found plentifully in the neighbourhood of scaly limestone, as flint is in the strata of chalk. What connexion there may be between these bodies, perhaps time will discover.

But flints and agates being always found in loose and single irregular nodules, and never in rocks, as the chert, is a circumstance very insufficient to establish a difference between them ; for there is agate near Constantinople running vein-like across the rock with its country, of the same hardness, and as fine and transparent as those other agates, which are found in round nodules at Deux Ponts. We must therefore content ourselves with this remark concerning flints, That they seem to be the only kind of stone hitherto known, of which a very large quantity has been formed in the shape of loose or separate nodules, each surrounded with its proper crust ; and that the matter which constitutes this crust, has been separated from the rest of the substance in like manner as sandiver, or glass gall, separates from, and swims upon glass during its vitrification ; tho' sometimes the formation of this crust may have been prevented by the too sudden hardening of the matter itself: I shall therefore take the liberty to call this matter of the crust, which sometimes is an indurated terre verte, by the name of *Agate-gall.*

Other species of stones, which are found in loose pieces, or nodules, except ores, and some sorts of stalactites, shew evidently by their cracks, angles, and irregular figures, that they have been torn from rocks, rolled about, and rubbed against one another in torrents, or by some other violent motions of water. That flints have originally been in a soft state, as I have mentioned, is easy to be seen in the Egyptian pebbles, which have impressions of small stones, sand, and sometimes perhaps grass, which however have not had any ingress into the very flint, but seem only to have forced the abovementioned agate gall or crust out of the way †.

SECT. LXIV.

G. Jasper, *Jaspis.*

    All the opaque flints are called by this name, whose texture resembles dry clay, and which have no other known quality, whereby

† The erroneous notion of the once soft state of stones, see discussed in my First Lecture on Fossils. D, C.

they

they may be diftinguifhed from other flints, except that they may be more eafily melted in the fire; and this quality perhaps may proceed from fome heterogeneous mixture, probably of iron.

1. Pure jafper, *Jafpis purus.*
Which by no means yet known can be decompounded.

    a. Green with red fpecks or dots, the *Heliotrope,* or blood-ftone, from Egypt.

    b. Green, from Bohemia.

    c. Red, *Diafpro roffo Italorum.*

    d. Yellow.

    e. Red with yellow fpots and veins, *Diafpro florido* of Sicily, Spain, and Conftantinople.

    f. Black, from Finland, and Neſkott, in the province of Jemtland.

SECT. LXV.

2. Jafper containing iron, *Jafpis martialis, Sinople.*

    A. Coarfe grained.

        a. Red and reddifh brown, *Sinople,* from the Hungarian gold mines.

    B. Steel grained, or fine grained.

        a. Reddifh brown, from Altenberg, in Saxony; looks like the red ochre or chalk ufed for drawing, and has partition veins, which are unctuous to the touch, like a fine clay, and other like kinds.

    C. Of a folid and fhining texture, like a flag.

        a. Liver-coloured, and

        b. Deep red. Both thefe are found at Lang-

Langbanſhyttan in the province of Wermeland, and at Sponwik in Norway.

c. Yellow, from Bohemia.

This laſt mentioned, when calcined,
is attracted by the load-ſtone, and being
aſſayed, yields 12 to 15 per cent. of
iron.

* Jaſper, when freſh broke, ſo nearly reſembles a bole of
the ſame colour, that it can only be diſtinguiſhed by its
hardneſs. In the pariſh of Orſa, in the province of Dalarne,
there is a red bole found in ſpaces like glands or kernèls, in
that ſort of ſandſtone from which grindſtones are cut; and
ſome miles diſtant, in the rocks at Serna, a red jaſper of the
ſame colour and textuie as the above bole, is found in a
much harder kind of ſandſtone. In other places jaſper is
found in ſuch unctuous clefts, as if they had contained unctuous clays; as pipe clays, and red chalk: and there are
likewiſe ſome jaſpers which imbibe water. May it not then
be ſuppoſed with ſome probability, that jaſper is an indurated bole, a reddle, or terre verte? That jaſper, as well as
theſe, conſiſts of clay and iron; though, by reaſon of its
being hardened, it becomes as difficult to extract theſe principles from it, as to reduce a ſmall quantity of ſcoriſied iron
to its metallic form, when melted with a large quantity of
ſlag or glaſs? That the ſame bole or clay, together with
another ſubſtance, perhaps lime, after being diſſolved by a
menſtruum, not yet determined, is ſufficient for the production of flint ſtone? and that ſo much of the bole as was
ſuperfluous, being ſeparated from the maſs, is found ad-
hering to the ſurface, or in the fiſſures, &c.

Thus one might imagine, that jaſper could eaſily be produced, and that the ſoft kinds might become harder by
length of time; but its particles cannot be ſuppoſed to approach nearer and nearer to one another during the hardening; nor can it be imagined, that the jaſper ſhould by that
means become of a finer texture. On the other hand, we
know extremely well, and have the experience of it every
where, that porphyry in the rocks decays into a white cruſt,
wherever it is expoſed to the air, although internally it remains very hard and black; for inſtance, at Klitten, in Elf-
dalen, in Sweden. From whence it may be ſuppoſed, that
water, which waſhes off the mouldered particles, muſt by
degrees collect them ſomewhere, and at length preſent us

with

SECT. LXVI.

H. Rhombic Quartz, *Spatum scintillans Felt-spatum* *.

This has its name from its figure †, but seems to be of the same substance as the jasper. I have not however ranked them together, for want of true marks to distinguish the different sorts of the flinty tribe from one another.

This kind is found,

1. Sparry.
   a. White.
   b. Reddish brown, occurs in the Swedish and also in the foreign granites.
   c. Pale yellow.
   d. Greenish.

   This last mentioned resembles very much the schorl or cockle-spar, (Sect. lxxiii.) but is neither so easy to melt in the fire, nor of so exact a figure.

2. Cristallised.

with them in form of an earth, which perhaps we do not know in that state. It may be asked, Whether this earth will be ductile as clay, or rough to the touch as powder of bricks? Perhaps Tripoly is produced in this manner.

* The Germans and other nations call this substance Feld-spat, that is feild or vague spar; a name very inadequate to its nature, as it touches no wise on the sparry class. I have therefore, in my Lectures, given it the English name of Rhombic Quartz, a name very significant, as the name quartz expresses its class; and the quality of it, of always breaking into angular (rhombic) fragments, is also expressed by the adjective of rhombic. D. C.

† What Cronstedt means, that the feld-spat has its name from its figure, is unintelligible. D. C.

A. In

A. In feparate or diftinct rhomboidal criftals, from the iron mine called Moff-grufvan, at Norbery in Weftmanland *.

SECT. LXVII.

Observation on the Siliceous Order.

The œconomical ufes of this order are not fo manifold as thofe of the Calcareous and Argillaceous claffes; however, moral reflections laid afide, it will be neceffary briefly to mention how far this order is confidered and employed in common life.

The Europeans have no farther trouble with the precious ftones, than either to cut them from their natural or rough figure, or to alter them, when they have been badly cut in the Indies; in which latter circumftance they are called *Labora:* and it may be obferved, that for cutting the ruby, fpinell, ballas, and chryfolite, the oil of

* This fpecies is very feldom found alone in form of veins, and ftill more rarely as conftituting the fubftance of whole mountains; but is generally mixed either with quartz and mica, as in the granites, or with jafper, having fome occafional concurring particles of quartz, cockle, and hornblende, as in the porphyry. If the rhombic quartz and jafper were of the fame fpecies, that fort of porphyry which is made up of thefe two bodies only, ought to be ranked among the jafpers, inftead of being placed with the Saxa in my Appendix, Sect. cclxvi. It is however obfervable on old monuments, which are expofed to the open air, that though the porphyry has decayed, and confequently loft its polifh, yet granite of the fame age, compofed for the moft part of rhombic quartz, has kept its luftre. This, however, does not contradict the poffibility of rhombic quartz being of the fame fubftance as the jafper; becaufe the calcareous fpar is found to bear the weather, and even fire, better than the limeftone.

vitriol

vitriol is required, inftead of any other liquid, to be mixed with the diamond powder.

If the petty princes in thofe parts of the Indies where precious ftones are found, have no other power nor riches proportionable to the value of thefe gems ; the reafon of it is as obvious as of the general weaknefs of thofe countries where gold and filver abound, viz. becaufe the inhabitants, placing a falfe confidence in the high value of their poffeffions, neglect ufeful manufactures and trades, which by degrees produces a general idlenefs and ignorance thro' the whole country.

On the other hand, perhaps fome countries might fafely improve their revenues by fuch traffic. In Saxony, for example, there might probably be other gems found, befides aquamarines and topazes ; or even a greater trade carried on with thefe than at prefent, without danger of bad confequences ; efpecially under the direction of a careful and prudent government.

The half precious ftones, fo called, or gems of lefs value, as the opal, the onyx, the chalcedony, the carnelian, and the coloured and uncoloured rock criftals, have been employed for ornaments and œconomical utenfils, in which the price of the workmanfhip greatly exceeds the intrinfic value of the ftones. The antients ufed to engrave concave or convex figures on them, which now-a-days are very highly valued, but often with lefs reafon than modern performances of the fame kind. Thefe ftones are worked by means of emery on plates of lead, copper, and tin, or with other inftruments ; but the common work on agates is performed at Oberftein, with grindftones, at a very cheap rate. When once fuch a

manufac-

manufactory is established in a country, it is
necessary to keep it up with much industry and
prudence, if we would wish it to surmount the
caprice of fashions ; since how much soever the
natural beauty of these stones seems to plead for
their pre-eminence, they will at some periods
unavoidably sink in the esteem of mankind, but
they will likewise often recover, and be restored
to their former value.

The grindstones at Oberstein are of a red colour, and of such particular texture, that they
neither admit of any polish, nor are they of too
loose a composition.

Most part of the flinty tribe is employed for
making glass, as the quartz, the flints and pebbles, and the quartzose sands. The quartz,
however, is the best ; and, if used in due proportion with respect to the alcali, there is no
danger of the glass being easily attacked by the
acids; as has sometimes happened with glass
made of other substances : for instance, of bottles filled with Rhenish and Moselle wines,
during the time of a voyage to China.

In the smelting of copper ores, quartz is used
to render the slag glassy, or to vitrify the iron ;
quartz being more useful than any other stone,
to regain or revive this metal.

The presence of the quartz in the rock-stones,
(Sect. cclxii.) and also in crucibles, and such vessels, contributes most of all to their power of
resisting the fire : it appears likewise probable,
that the quartzose matter makes the grind and
whetstones fit for their intended purposes.

SECT.

SECT. LXVIII.

The THIRD ORDER.

The Garnet Kind, *Terræ Granateæ*.

The matter compofing the fubftance of garnet, and fchorl or cockle, except that fmall portion which is metallic, does in its indurated ftate refemble the filiceous tribe, fo far as relates to external appearance and hardnefs; and therefore I would willingly have followed the opinion commonly received, of confidering thefe two fubftances as arifing from one another, if I had not been perfuaded to the contrary by the following qualities of the garnet.

1. It is more fufible, in proportion as it contains lefs metallic matter, and is more tranfparent or glaffy in its texture; which is quite contrary to the filiceous kind.

2. This is the reafon, perhaps, why the garnet, mixed with the falt of kelp, may on a piece of charcoal be converted to a glafs by the blow-pipe, which cannot be done with the flints: and,

3. Why the moft tranfparent garnet may, without any addition, be brought to a black opaque flag by the fame means.

4. It is never, fo far as is hitherto known, found pure, or without fome mixture of metal; and efpecially iron, which may be extracted by the common methods.

5. The garnet matter, during the criftallifation, has either been formed in fmall detached quantities, or elfe has had the power of fhooting into criftals, though clofely

clofely confined in different fubftances : fince garnets are generally found difperfed in other folid ftones, and oftentimes in the harder ones, fuch as quartz and chert *.

SECT. LXIX.

1. Garnet, *Granatus.*
Which is a heavy and hard kind of ftone, criftallifing in form of polygonal balls, and is moftly of a red, or reddifh brown colour.

A. Garnet mixed with iron, *Granatus martialis.*

---

* It is certain, that the metallic calces being mixed with other earthy fubftances, make great alteration in refpect to their fufibility; and we know from experience, that the prefence of iron in the argillaceous, and moft particularly in the micaceous kinds, renders them fufible ; however, though there may be good reafons for confidering the garnet as a quartz impregnated with iron ; yet as quartz becomes lefs fufible by any addition of iron, of which the Swedifh *Torrften,* (Sect. ccxiii. in the note) a martial ore, commonly mixed with quartz, is an inftance ; and as even the fpecies of quartz naturally mixed with that metal, (Sect. liii. *A.*) are far lefs eafily fufible than the garnet ; it might perhaps be better to call the garnet a ftone of a different order, until by fufficient experiments we may have reafon to reduce the number of the earths. Though if we fhould ever arrive at an exact method of claffing in this refpect, perhaps the œconomical ufe of thefe bodies will rather require fuch a diftribution of them as fhall more regard their prefent exiftence, than that which they have been originally derived from. The garnet earth, fo far as I know, is not yet found, but in an indurated ftate; and, as fuch, it is divided into the garnet, and into fhirl or cockle, and that in regard to the figure of their criftals, more than any thing elfe : though their colour has alfo had fome fhare in this divifion. I have here followed cuftom, which, perhaps, may have fome reafon, however ill founded it be.

1. Coarfe

1. Coarfe grained garnet ftones, without any particular figure, *Granatus particulis granulatis figura indeterminata* ; in Swedifh called *Granat-berg* ; in German *Granatftein*.

   a. Reddifh brown garnet, found in the mine called Granat-Skierpningen, at Nya Kopparberg, in the province of Weftmanland.

   b. Whitifh yellow, from Torrakeberget, in the parifh of Gorborn, in Wermeland.

   c. Pale yellow, from Sikfioberget, and Vefterfilfverberget, in Kopparbergflan, in Sweden.

2. Cryftallifed garnet, *Granatus cryftallifatus*.

   a. Black, from Swappawari, in Lapland.

   b. Red, femi-tranfparent, and cracked, from Engfo, at the Lac Malaren, in Weftmanland.

   Tranfparent, *Granatus gemma*.

   c. Reddifh yellow tranfparent, the jacinth, or hyacinth, *Hyacintus gemma*, from Greenland, and Bergen's Stift in Norway.

   I am not certain whether the oriental jacinth and that from Siberia are of the fame kind; but this garnet from Greenland is by the jewellers fold as a jacinth.

   d. Reddifh brown, from Kallmora and Stripas, at Norbery, in Weftmanland.

   e. Green, from Eibenftock in Saxony, and Gellebeck, in Norway.

   f. Yellowifh green, from Gellebeck.

SECT.

SECT. LXX.

B. Garnet mixed with iron and tin, *Granatus crocis Martis et Jovis mixtus.*
1. Coarfe grained, without any particular figure, *Granatus particulis granulatis figura indeterminata.*
   a. Blackifh brown, from Moren, at Weftanfors, in Weftmanland.
2. Cryftallifed.
   a. Blackifh brown, from Moren.
   b. Light green or white, from Gokum, at Dannemora, in the province of Upland.
   The Bergs-radets, or mine-mafters, Mr. Brandt and Mr. Rinman have publifhed fome experiments on this kind of garnet, in the Memoirs of the Royal Academy of Sciences at Stockholm.

SECT. LXXI.

C. Garnet mixed with iron and lead, *Granatus calcibus Martis et Saturni mixtus.*
1. Cryftallifed.
   a. A reddifh brown, from Arfet, in the parifh of Froderyd, in the province of Smoland.
   This was difcovered, and accurately examined by the Bergs-radet Mr. Von Swab*.

---

* When any of the garnet kind is to be tried for its containing metal, the iron ought to be melted out of it by the common procefs; and if the garnet, at the fame time, contains both tin and lead, thefe two metals are likewife included in the iron: however, they may be extracted out of the iron,

by

SECT. LXXII.

2- Cockle or Shirl, *Bafaltes*; *Corneus cryftallifatus Wallerii*; *Stannum cryftallis columnari-
bus nigris Linnæi.*

Is a heavy and hard kind of ftone, which
fhoots into cryftals of a prifmatical figure,
and whofe chief colours are black or green.
Its fpecific gravity is the fame as the garnets,
viz. between 3000 and 3400, though always proportionable to their different folidity.

A. Cockle, or fhirl, mixed with iron, *Bafaltes
martialis.*

1. Coarfe, without any determined figure,
*Bafaltes particulis palpabilibus figura in-
determinata.*

a. Green, found in moft of the Swedifh
iron mines.

by expofing it to a heat augmented by degrees, becaufe then
the tin and lead fweat out in form of drops, almoft pure,
though always fomewhat mixed with iron.

The criftallifations of the garnets are fo far different from
one another, that fome have a greater and fome a lefs number of facets, or fides; but this has no relation or dependence
either on their contained metals, their colour, or their
tranfparency; wherefore, in order to avoid a prolixity,
which is unneceffary, I have omitted fuch varieties, and only
obferved that they are round or fpherical, with facets. Befides, there is not yet difcovered any figure amongft them
which is abfolutely particular and remarkable, for the
granatus dodecaedros ex rhombis Linnæi are difperfed every
where in the rocks at Kongsberg, in Norway.

SECT.

SECT. LXXIII.

2. Sparry, *Bafaltes fpatofus.*
    a. Deep green, (the mother of the eme-
    ralds) from Egypt *.
    b. Pale green, from Wefterfilfverberget,
    and Hagge, at Norberke, Linbaftmoren,
    at Grangierde, in the province of Weft-
    manland, &c.
    c. White, from Silf-udden, at Wefter-
    filfverberget, Pargas in Finland, the
    lime-rocks at Lillkyrkie, in the province
    of Nerike, &c.

    This occurs very frequently in the
    fcaly lime-ftones, and its colour changes
    from deep green to white, in proportion
    as it contains more or lefs of iron.

SECT. LXXIV.

3. Fibrous, *Bafaltes particulis fibrofis* ; Striated
    cockle, or fhirl : it looks like fibres, or
    threads made of glafs.
    A. Of parallel fibres, *Bafaltes fibris paral-
    lellis.*
    a. Black, from Guftavsberg, in the pro-
    vince of Jemtland, the ifland Uto, in
    the Lake Malaren, &c.
    b. Green, in moft of the Swedifh iron
    mines.
    c. White, from Wefterfilfverberget, in
    the province of Weftmanland, Lill-
    kyrkie in Nerike, and Pargas in Fin-
    land.

* The Plafma, or mother of the emerald of authors, is a fine
pellucid true quartz, of a green or emerald colour, not a ftone
of this kind. D. C.

G   *B.* Of

B. Of concentrated fibres, *Bafaltes fibris concentratis*; The ftarred cockle, or fhirl, from its fibres being laid ftellarwife.

a. Blackifh green, from Salberg, in Weftmanland, where, being found together with a fteel grained lead ore, the whole is called, *gran-ris-malm*, or pine-ore, from its refemblance to the branches of that tree. This kind of cockle is alfo found at Uto, in Malaren.

b. Light green, from Kerrbo, at Skinfkatteberg, in Weftmanland.

c. White, at Lillkyrkie in Nerike, Wefterfilfverberget in Weftmanland, and Pargas in Finland.

S E C T. LXXV.

4. Cryftallifed cockle, or fhirl, *Bafaltes cryftallifatus.*

a. Black, from France, Yxfio at Nya Kopparberg, in the province of Weftmanland, Umea in Lapland, Ofterbottn in Sweden.

---

* To this fpecies of cockle, or fhirl, belong moft of thofe fubftances called *imperfect afbefti*; and as the cockle perfectly refembles a flag from an iron furnace, both in regard to its metallic contents, and its glaffy texture, it is no wonder that it is not foft enough to be taken for an afbeftus. It has however, only for the fake of its ftructure, been ranked among the afbefti; and it is furprifing, that the fibrous gypfum, from Andrarum, in the province of Skone, has efcaped being on the fame account confounded with them. The ftriated cockle, or fhirl, compared to the afbefti, is of a fhining and angular furface (though this fometimes requires the aid of the magnifying-glafs to be difcovered) always fomewhat tranfparent, and is pretty eafily brought to a glafs with the blow-pipe, without being confumed, as the pure afbefti feem to be. (See Afbefti, Sect. cii.)

b. Deep

b. Deep green, from Salberg in Weftmanland.

c. Light green, from Enighets-grufvan at
Norberg, in Weftmanland,

d. Reddifh brown, from Sorwik, at Grengie
in Weftmanland, and Glanfhammar, in
the province of Nerike.

The Tauffstein, from Bafil, is of this
colour, and confifts of two hexagonal
criftals of cockle grown together in form
of a crofs : this the Roman Catholics wear
as an amulet, and is called in Latin, *lapis
crucifer*, or the crofs-ftone * †.

* It is not impoffible, that there may be fome kinds of
cockles, or fhirls, which, befides iron, alfo contain tin or
lead, as the garnets : but I am not quite convinced of it ;
though I have been told, that lead has been melted out of a
cockle, from Rodbeck's Eng, at Umea, in Lapland ; and it
feems likewife very probable, that the cockles which are
found in the Englifh tin mines, may contain fome tin. There
are fome criftals of cockle found, which are fufible to a
greater degree than any fort of ftone whatfoever : thefe are
always of a glaffy texture, and femi-tranfparent.

The figure of the cockle criftals is uncertain, but always
prifmatical : the cockle from Yxfio, at Nya Kopparberg, is
quadrangular ; the French kind has nine fides, or planes, and
the Tauffftein is hexagonal ‖.

† The crofs-ftone is compofed of two claffes, for the bafis
I make a fluor, (See my Lectures) and the croffes on it I agree
with our author are accretions of fhirl, or cockle. He
denominates it, the *Bafler Tauffstein.* Whether he means by
Bafler the city of Bafil, as I have put it, or whether he means
Befler, the author who firft defcribed it, I cannot tell ; but certainly it is not found at or near Bafil, being, as far as I know,
a local foffil, namely, of St. John de Compoftella, in Anda-
lufia in Spain. D. C.

‖ The name Cockle for thefe fubftances is an old Cornifh mineral name ;
but is alfo given fometimes to other very different matters. The name Shirl
I have now adopted in Englifh, from the common German mineral term.

We have not in England any great quantity of fpecies of cockles ; the
chief are found in the tin mines of Cornwall, and I have feen fome fine
criftallifed kinds from Scotland.

The

SECT. LXXVI.

OBSERVATION on the GARNET KIND.

When this kind contains so much of iron as renders it profitable to be worked, it is confidered as a good iron ore, and no notice is taken of its natural character, in the fame manner as is done with clays and jafpers that contain iron : for the richnefs of metal in thefe rifes in a gradual progreffion, until they acquire the colour and appearance of the iron itfelf.

Thus a kind of garnet is melted in a furnace, not far from Eibenftock, in Saxony, and the fame fpecies is found, and might alfo be employed at Moren, in Weftmanland. Jafpers are for this purpofe melted in Hungary, and clays in England; but as the greateft part of the garnet kind contains fo little iron as to yield only between fix and twelve per cent. which is too poor to be worked any where in the world as a profitable iron ore, the reft and the greateft part of it being a mere earth, it muft in a natural hiftory be confidered and ranked among the earths.

The tin grains fhould have got a place in this order, 1. If I had known any of them to contain tin in fo fmall a portion as five per cent. as this quantity of tin is the moft that ever can be obtained from the garnets; 2. If it was proved that a calx of iron always was mixed with it, as in the garnet; and, 3. If I did not believe that the

---

The Englifh mineral name of *Coll*, has been ufed by fome authors as fynonymous with cockles, and is even confounded together at the mines; but the Call, definitely fpeaking, is the fubftance called Wolffram by the Germans, &c.

Garnets, though fmall, are often found in micaceous ftones in England; but extreme good garnets are found in great plenty alfo in like ftones in Scotland. D. C.

tin

tin calx might by itfelf take a fpherical polygonal figure, at its induration, as well as the garnet. The white tin grains, (Sect. ccx.) out of which no tin, but only iron, is to be got, might with more reafon be placed here, if it was not fo exceffively refractory in the fire, and if it did not, at laft, melted either by itfelf, or with borax, give a clear and colourlefs glafs, contrary to what the garnet does, which difference arifes from the different fufibility of thefe two fubftances.

The garnet and cockle are not yet known to me in form of an earth or clay, taken in the common idea we have of thofe bodies. It is true, that there is a bole found at Swappawari, in Lapland, which has the fame figure as the garnet; and the hornblende, (Sect. lxxxviii.) which is fomewhat harder than this bole, has often the appearance of a cockle. We cannot, however, do more than problematically fuppofe them to be the neareft related to the garnet kind, as we have not yet difcovered a method how to feparate earths from the contained metals, without deftroying their natural form, and efpecially from iron, when it is fo ftrongly united with them, as if it had a part in their formation itfelf.

SECT. LXXVII.

The FOURTH ORDER.

The Argillaceous Kind, *Argillaceæ.*

The principal character whereby thefe may be diftinguifhed from other earths, is, that they harden in the fire, and are compounded of very minute particles, by which they receive a dead or dull appearance when broken.

G 3

More-

Moreover, there are some of this order which grow soft in water, and, when only moistened, become ductile and tenacious : these are commonly called clays. Some crack in the water, after having imbibed a sufficient quantity of it, but do not grow softer in it, and are therefore in the first degree of induration : some imbibe the water, but do not crack or fall to pieces ; these are yet more indurated : and finally, some there are, in which the water has no ingress at all. Thus, by following the successive gradation of induration of a substance, which throughout all these circumstances is easily discovered to be the same, one may with great reason conclude, that the hardness of the jasper may perhaps be the last degree of hardness, and that this stone consequently consists of an argillaceous substance, (Sect. lxv.) that already possesses a quality which the other clays cannot acquire but in the fire ; having, besides, the same effect as the boles (Sect. lxxxvi.) when melted in the fire together with calcareous or other earths.

SECT. LXXVIII.

A. Porcellain Clay, *Terra Porcellanea,* vulgò *Argilla Apyra.*

Is very refractory in the fire, and cannot in any common strong fire be brought into fusion any farther than to acquire a tenacious softness, without losing its form : it becomes then of a dim shining appearance and solid texture, when it is broke ; strikes fire with steel ; and has consequently the best qualities required, as a substance whereof vessels capable of resisting a melt-

ing

ing and boiling heat, and of holding ſalts and acids, can be made. It is found,

1. Pure, *Pura.*
   A. Diffuſible in water.
   1. Coherent and dry.
      a. White, from Japan.

      I have ſeen a root of a tree changed into this clay. (Appendix, Section cclxxxiii.)
   2. Friable and dry.
      a. White, is found in clefts of rocks at Weſterſilfverberget in Weſtmanland, and between the coal, in the coal-pits at Boſerup, in the province of Skone *.

2. Mixed with phlogiſton, and a very ſmall quantity of inſeparable heterogeneous ſub-ſtances, *Terra porcellanea phlogiſto aliiſque heterogeneis minimâ portione mixta.* Of theſe are,
   A. Diffuſible in water,
      a. White and fat pipe clay, from Cologne and Maeſtricht.

      Leſs unctuous is found in ſmall fiſ-ſures in a vein of *lapis ollaris*, at Swart-wik, in the pariſh of Swerdſio, in the province of Dalarne.
      b. Of a pearl colour, from Maeſtricht.
      c. Bluiſh grey, *La belle terre glaiſe* from Montmartre, near Paris, in France.
      d. Grey, France, Heſſe, Boſempin, Skone.

---

* Theſe may be called pure, ſince after being burnt, they are quite white, though they have been expoſed to a quick melting heat ; and it may be queried, if all ſuch clays muſt not be ſomewhat harſh, or at leaſt not unctuous to the touch.

e. Black,

e. Black, *La terre noire*, at Montmartre.
f. Violet, alfo from Montmartre *.

SECT. LXXIX.

B. Indurated, *Indurata.*

Is commonly unctuous to the touch, and more or lefs difficult to be cut or turned, in proportion to its different degrees of hardnefs; is not diffufible in water, grows hard, and is very refractory in the fire; pounded and mixed with water, it will not eafily cohere in a pafte : however, if it is managed with care, it may be baked in the fire to a mafs, which, being broke, fhews a dull and po-

---

* Thefe contain a phlogifton, which is difcovered by expofing them to a quick and ftrong fire, in which they become quite black interiorly, affuming the appearance of the common flints, not only in regard to colour, but alfo in regard to hardnefs: but if heated by degrees, they are firft white, and afterwards of a pearl colour. The fatter they feem to be, which may be judged both by their feeling fmooth and unctuous, and by their fhining, when fcraped with the nail, they contain a larger quantity of the inflammable principle. It is difficult to determine, whether this ftrongly adherent phlogifton is the caufe of the above-mentioned pearl colour, or prevents them from being burnt white in a ftrong fire : yet no heterogeneous fubftance can be extracted from them, except fand, which may be feparated from fome, by means of water, but which fand does not make out any of the conftituent parts of the clays. If they be boiled in aqua regis, in order to extract any iron, they are found to lofe their vifcofity. In the lefs unctuous clays, I have found pure quartz in greater and fmaller grains; but ftill I would not venture to affert, that one is produced from the other, according to the rule I have laid down in Sect. ix. 1. I have likewife found this fort, upon certain occafions, attract the phlogifton in the fire. Thefe remarks may ferve as hints for the lefs experienced, who have a mind to examine thofe clays, which are of fo great confequence for their œconomical ufes.

rous

rous texture. It takes for the moſt part, and without much labour, a fine poliſh. It is found,

1. Compact and ſoft, *Particulis impalpabilibus mollis*; Smectis, Briançon, or French chalk.
   a. White, from the Lands End, in Cornwall.
   b. Yellow.
   c. Red and white, Land's End: the Soap Earth, Switzerland. It looks like Caſtile ſoap.

SECT. LXXX.

2. Solid and compact, *Particulis impalpabilibus ſolida*; Steatites, and alſo Soap Rock.
   a. White, or light green, from Riſver, in Norway, Bareuth, and Sikſioberget, at Norberke, in Weſtmanland.
   b. Deep green, from Salberg, in Weſtmanland, Swartwik, in Dalarne, Jonuſwando, in Lapland, Salviſto, at Tamela, in Finland, &c.
   c. Yellow, from Juthyllen, at Salberg, Torråkeberget, at Goſborn, in the province of Vermeland, and China*.

* It is a very difficult matter to ſpecify all the varieties of the ſoap-ſtones, in regard to their hardneſs or ſoftneſs, ſince they cannot be compared with any ſtandard meaſure. Thoſe from Riſver, Sikſioberg, and China, are a great deal harder and more ſolid than the Engliſh kind, from the Land's End, which breaks between the fingers; but are ſoft in compariſon to that from Salberg, which is there called *Serpentine*, although both theſe varieties may indiſcriminately be made uſe of for cutting and turning. The ſoft ones, however, are not ſo apt to crack, when they are worked, as the harder. But

SECT. LXXXI.

3. Solid, and of vifible particles, *Solida particulis majoribus*; Serpentine ftone, *Lapis Serpentinus.*

1. Of fibrous and coherent particles, *Lapis ferpentinus fibrofus.*

This is compofed, as it were, of fibres, and might therefore be confounded with the afbeftus, if its fibres did not cohere fo clofely with one another, as not to be feen when the ftone is cut and polifhed. The fibres themfelves are large, and feem as if they were twifted.

q. Deep green.

Is fold for the *lapis nephriticus,* and is dug at fome unknown place in Germany.

b. Light green, from Skienfhyttan, in Weftmanland; is ufed by the plate-fmiths, inftead of the French chalk.

SECT. LXXXII.

2. Fine grained Serpentine ftone, *Serpentinus particulis granulatis* : the Zoeblitz Serpentine.

But none of thefe varieties is found in the rock, without being interfperfed with the unctuous clefts. When they are too many, too clofe to one another, and make the ftone unfit for ufe, they are in this cafe called by the Swedifh miners, *Skiolige*; and of this kind is a great quantity found at Salberg and Swartwik. Moft part of the foap-rock, which is found in Sweden, is likewife mixed with glimmer or mica, and then it is called *Telgften,* that is, *Ollaris,* Sect, cclxv.

a. Black.

a. Black.

b. Deep green.

c. Light green.

d. Red.

e. Bluiſh grey.

f. White. Theſe colours are all mixed together in the ſerpentine ſtone, from Zoeblitz, but the green is the moſt predominant colour.

SECT. LXXXIII.

3. Mixed with iron, *Terra porcellanea marte
mixta.* This is

A. Diffuſible in water.

    a. Red, *la terre rouge,* from Montmartre,
and China.

    Some of the bricks which are imported
from ſome certain places in Germany,
ſeem to be made of this kind.

B. Indurated.

    1. Martial ſoap earth, *Creta Brianzonica
martialis.*

      a. Red, from Jarſberg, in Norway.
It is likewiſe mixed with ſome calcareous matter.

    2. Martial ſoap rock, *Steatites martialis.*

      a. Black, from Sundborn, in Dalarne,
Torrakeberget, in Wermeland, Offerdal, in Jemtland.

      b. Red, from Siljejord in Telemarken,
in Norway *.

---

* Since the iron renders the ſo called refractory clays, as
well as other clays, eaſier fuſible than they really are by
themſelves ; it might be queried, how it can be determined,
of what ſpecies of argillaceous matter theſe conſiſt ? To this
it

SECT. LXXXIV·

B. Stone Marrow, *Lithomarga* : *Keffekil* of the
Tartars.

I have given this name to a kind of clay,
which,

1. When dry, is as fat and flippery as
   foap : but,
2. Is not wholly diffufible in water, in which
   it only falls to pieces; either in larger bits,
   or refembles a curd-like mafs.
3. In the fire it eafily melts to a white or
   reddifh frothy flag, therefore confequently
   is of a larger volume than the clay was
   before being fufed.
4. It breaks into irregular fcaly pieces.
   A. Of coarfe particles : Coarfe Stone Marrow.
   a. Grey, from Ofmundfberget, in the
   parifh of Rettwik, in Dalarne, and
   is there called *walklera*, that is, fuller's
   earth. It is mentioned in an account
   of Ofmundfberget, publifhed in the
   Tranfactions of the Academy of
   Sciences at Stockholm, in the year
   1739, by the Berg's-radet, or mine-
   mafter, Mr Tilas.
   b. Whitifh yellow, from the Crim
   Tartary, where it is called Keffekil,

---

it is anfwered, That they are found together in the fame beds
with the porcelain clay : that they have all the fame external figns, and differ from it only in the colour, being red,
brown, or black, in regard to the contained metal : that
they are more refractory in the fire than any other martial
clay ; and that, though they may be reduced fo as to refemble
a black or iron-coloured flag, they yet retain their form.

and

and is faid to be ufed for wafhing inftead of foap.

B. Of very fine particles: Fine Stone Marrow.

a. Yellowifh brown, *Terra Lemnia*.

Is of a fhining texture, falls to pieces in the water with a crackling noife; it is more indurated than the precedent, but has otherwife the fame qualities *.

SECT. LXXXV.

C. Bole, *Bolus.*

Is a fine and denfe clay of various colours, containing a great quantity of iron, which makes it impoffible to know the natural and fpecifical qualities of the bole itfelf, by any eafy method hitherto in ufe. It is not eafily foftened in water, contrary to what the porcelain and the common clays are (*A. & E.*), but either falls to pieces in form of fmall grains, or repels the water, and cannot be made ductile. In the fire it grows black, and is then attracted by the load-ftone.

---

* This cannot properly be called a *fuller's earth*, fince it neither is of that kind ufed in the fulling bufinefs, nor is likely to be applicable to it ‡. It is, befides, a very fcarce clay. It is not found indurated, fo far as I know; and if it fhould at any time be difcovered, it will be neceffary to examine, if it is not a Zeolites (or the eighth order), or at leaft very nearly approaching to it, in regard to the effects both undergo in the fire.

‡ As the beft fort of Fuller's Earth did not come into our author's hands, it is no wonder that he excludes it from its due place. The true Fuller's Earth of England is exactly like the ftone marrow in all the above-mentioned properties; and in regard to the texture and colour, it comes neareft to the above-defcribed coarfe ftone marrow.   E.

SECT.

SECT. LXXXVI.

1. Loofe and friable boles, or thofe which fall to a powder in water.

a. Flefh-coloured bole, from Kriftiersberg, at Nya Kopparberg, in Weftmanland.

b. Red.

 1. Fine, *Bolus Armenus.*

 2. Coarfe, *Bolus communis officinalis,* from the fand-ftone quarries at Orfa, in the province of Dalarne.

 3. Hard, *Terra rubrica.*

c. Green, *Terre verte.*

 1. Fine, from Italy.

 2. Coarfe, from Stenftorp, in the province of Weftergottland.

d. Bluifh grey, from Stollberget, in Kopparbergflan, in Sweden.

 Is ductile as long as it is in the rock, but even then repels the water; it contains forty per cent. of iron; which metal being melted out of it in a clofe veffel, the iron cryftallifes on its furface.

e. Grey.

 1. Cryftallifed in a fpherical polygonal figure: from Swappawari, in Lapland.

 2. Of an undetermined figure, from Grengerberget, in Weftmanland *.

‡ At the time when the *terræ figillatæ,* or fealed earths, were in general ufe, the druggifts endeavoured to have them of all colours; and for that reafon they took all forts of clays and fealed them; not alone the natural ones, but likewife fuch as had been coloured by art, or had been mixed with *magnefia alba officinalis,* or other things, were afterwards vended for true boles; and for this reafon the fpecies of

boles

SECT. LXXXVII.

2. Indurated Bole, *Bolus indurata.*
A. Of no vifible particles, *Particulis impalpabilibus.*

This occurs very often in form of flate, or layers in the earth, and then is made ufe of as an iron ore. However, it has ufually been confidered more in regard to its texture, than to its conftituent parts, and has been called flate, in common with feveral other earths, which are found to have the fame texture.

a. Reddifh brown, from England †.
b. Grey, from Coalbrookdale, in Shropfhire, and moft collieries of England.

SECT. LXXXVIII.

B. Of fcaly particles, *Particulis fquamofis:* The hornblende of the Swedes.

boles is ftill thought to comprehend fo many varieties. Thus the Cologne clay (Sect. lxxviii.) is by the druggifts ranked among the white fealed earths, and is called a *white bole:* and this fame clay is by the Swedifh potters called *Englesk jord,* or Englifh earth; and by the tobacco-pipe makers, *Pip-lera,* or pipe clay, &c. which fhews how great a confufion there muft enfue, if the knowledge of thefe bodies was not founded upon a furer ground than the colour, figure, and names invented by common mechanics. Since the moft part of thefe *terræ figillatæ,* or fealed earths, are found to contain iron, I conclude, that the bole muft be a martial clay; and, as fuch, it feems to be more fit for medical ufes than other clays, if any dead earth muft be ufed internally, when there is fuch an abundance of finer fubftances.

† In moft collieries between the feams of coal, as at Hannam, in Kingfwood, near Briftol, Blanavon, in Monmouthfhire, &c. D. C.

Is

Is diftinguifhed from the martial glimmer, or mica, (Seft. xcv.) by the fcales
being lefs fhining, thicker, and reftan-
gular.

a. Black. This, when rubbed fine, gives a
green powder.

b. Greenifh.

Both thefe, particularly the black, are
found every where in Sweden among the
iron ores, and in the Grunften (Seftion
cclxix *).

S E C T. LXXXIX.

D. Tripoli, *Terra Tripolitana.*

Is known by its quality of rubbing hard
bodies, and making their furfaces to fhine,
the particles of the tripoli being fo fine, as
to leave even no fcratches on the furface.
This effeft, which is called polifhing, may
likewife be effefted by other fine clays, when
they have been burnt a little. The tripoli
grows fomewhat harder in the fire, and is very
refraftory : it is with difficulty diffolved by
borax, and ftill with greater difficulty by the
microcofmic falt : it becomes white when it is
heated : when crude, it imbibes water, but
is not diffufible in it : it taftes like common
chalk, and is rough or fandy between the
teeth, although no fand can by any means be
feparated from it. It has no quality common

---

* The hornblende grows hard in the fire, which is the reafon why it is ranked here among the clays, though in all its
other qualities it much refembles the cockle or fhirl. (Seftion
lxxii.) E.

with

with any other kind of earth, by which it might be confidered as a variety of any other. That which is here defcribed, is of a yellow colour, and is fold by druggifts, who do not know where it is found *.

SECT. XC.

E. Common Clay, or Brick Clay, *Argilla communis ; vulgaris Plaftica.*

This kind may be diftinguifhed from the other clays, by the following qualities.

1. In the fire it acquires a red colour, more or lefs deep.
2. It melts pretty eafily into a greenifh glafs.
3. It contains a fmall quantity of iron and of the vitriolic acid, by which the preceding effects are produced.

It is found,

A. Diffufible in water.
  1. Pure.
    a. Red clay, from Kinnekulle, in the province of Weftergottland.
    b. Flefh coloured, or pale red, is found on the plains between Wefteras and Sala, in the province of Weftmanland.
    c. Grey, in the corn-fields in the province of Upland.

---

* I have got of this kind of tripoli from Scotland, which has been lately difcovered there. But the rotten-ftone, fo called, is another fort found in England, viz. in Derbyfhire. It is in common ufe here in England among workmen for all forts of finer grinding and polifhing, and is alfo fometimes ufed by lapidaries for cutting of ftones, &c. D. C.

H   *d.* Blue

d. Blue, is very common in Sweden, in the provinces bordering upon the Baltic.

e. White, is found in the woody parts of Sodermanland, Dalarne, and of other provinces. It is often found in a flaty form, with fine sand between its strata. It is not easy to be baked in the fire : when it is burnt, it is of a pale red colour, and is more fusible than the preceding ones.

f. Fermenting clay, *Argilla intumescens.*

This is very like the preceding (*e*), as to the external appearance, and other qualities ; but when they are both found in the same place, which is not uncommon in several of our mine countries, they seem to be different in regard to the fermenting quality of this variety. This fermentation cannot be the effect of the sand mixed with it, because sand is found in them both : and besides, this kind ferments in the same manner when it is mixed with gravel or stones; and then it ferments later in the spring than the other, since by the stones, perhaps, the frost is longer retained in it.

2. Mixed with lime, see Marle, Section XXV.

SECT.

SECT. XCI.

B. Indurated.
1. Pure.
   a. Grey flaty.
   b. Red flaty, from Kinnekulle, in the province of Weftergottland.
2. Mixed with phlogifton, and a great deal of the vitriolic acid. See Alum Ores, Sect. cxxiv.
3. Mixed with lime. See Lime, Sect. xxviii *.

* It is probable, although it is not eafily demonftrable, that the common clay, and efpecially the blue, grey, and pale red, which are the foils of our plains and dales bordering upon lakes, has its origin from mud, and that the mud owes its exiftence to vegetables ; confequently that thefe varieties of clay are nothing elfe than a.mould, or *humus ater*, fomewhat altered by means of water, and by length of time. The following circumftances contribute greatly to confirm this opinion, viz. that a great quantity of fea-plants rot every year in the lakes, and are changed into mud ; that very little, however, of this mud is feen upon the fhores after the water is dried in fummer-time ; and that the clay begins where the mud ceafes. Concerning the turf, or peat, it is to be obferved, that this is not always produced from vegetables growing upon the very fame fpot where it is cut, but from fuch vegetables as have been thrown together from other places : for in what other manner could hazle-nuts occur in the turf-moors, in places where no hazle-trees grow, even at a diftance of many miles ? not to mention other inftances of the fame nature. Secondly, the turf, or peat, is cut in humid and low marfhes, which are not conftantly covered with water, as on the banks of lakes over-grown with grafs. If the origin of turf was any other than here mentioned, there ought to be turf found inftead of mud at the bottom of lakes where there is plenty of grafs.

The quantity of iron, and of the vitriolic acid contained in this clay, would perhaps not be found greater than to anfwer in proportion to the quantity of each of thefe fubftances, that enters into the compofition of vegetables, whilft growing, if there were any poffibility of making the comparifon. Mean

while

SECT. XCII.

OBSERVATION on CLAYS in general.

Thofe who have taken upon themfelves to examine the mineral bodies according to the principles upon which this Syftem is built, will readily, I hope, excufe thofe faults which may have been committed in claffing the clays; becaufe they muft well know, not only how difficult it is to procure a number of different varieties of this order in their natural ftate, which have not been previoufly wafhed or prepared for

while I have in dry fummers obferved on the fea-fhore, that a perfect iron vitriol has been growing out of the mud, clays, and vegetables not yet rotted, which has been thrown up there together.

When this opinion is once proved to be true, one may venture to go farther, and endeavour by obfervations and experiments to prove likewife, that in the fubverfions or changes that the earth has more than once fuffered in every part of it, and in which water has contributed the moft to carry off the particles, and to change the ftrata, the clay has been gathered together, and lodged in beds together with other fubftances. Some of thefe ftrata have afterwards been indurated, by which means they are turned into the above flaty and limy clays; and when they have been mixed with a great quantity of vegetables, and of the inflammable fubftance, they may in length of time be changed into pit-coal: but when they have been mixed with lefs phlogifton, and a great quantity of the vitriolic acid, they conftitute the alum ores, &c.

Others of thofe ftrata, which are not yet hardened, prove ftill, by their being fet or divided with fome feparating veins of fand, that they have been formed in the fame manner as the fettlings or fediments of ftamping mills, and may perhaps, through edulcoration in water, or through age, have loft their fertility, fince they never are fo good to improve lands with, as thofe ftra'a which are fuppofed to be of a more recent formation, fuch as *b. c. d.*

ufe,

ufe, as the fealed earths, &c. but alfo that it is
no eafy matter diftinctly to defcribe fome little
circumftances that occur to the eye, both in their
natural ftate, and during the experiments. Befides, they cannot but remember, that the pro-
greffional degrees both of hardnefs, and of the
quantity of mixed heterogeneous bodies, efpecially iron, produce a number of imperceptible
differences between them, in regard to colour and
effects; fo that they cannot with due precifion be
feparated and divided into their true genera, fpecies, and varieties, before fome more evident
differences between them may, by repeated experiments, and perhaps by proceffes yet unknown, be difcovered. In examining the clays,
one ought carefully to obferve the different degrees of fire due to each kind: for without this
knowledge they can never be employed to any
real ufe in common life. Next to this, there is
another point equally neceffary to be taken notice of, that is, the manner of working the clays,
which is often different in different kinds, and
which, not lefs than the different degrees of fire,
is productive of different effects; and therefore,
if both thefe circumftances are not at the fame
time exactly defcribed, it is as wrong to affert
with fome authors, that a refractory clay does never
crack in the fire, as it is deceiving to pretend that
the fame clay does never imbibe the water, when
it has been baked. Hence comes that great difference in regard both to appearances and qualities, between a tobacco-pipe, which is very little
baked, and a jar from Waldenburg, between a
common brick and the other fort called a water
clinkert.

H 3

The

The ufe of clays, in common life, is more extenfive than I have been able to inform myfelf of; for which reafon I will only mention fome particulars relating to it.

The porcelane clay is employed to make veffels which have that quality already mentioned (Sect. lxxviii.). I make no doubt but it enters into the compofition for making the fine porcelane ware at fome places; at leaft veffels are prepared from it of the fame goodnefs in every refpect: and there are likewife fome varieties of this clay, which become quite white in the fire, a quality which is efteemed the moft valuable in the fine China ware.

The indurated porcelane clay cannot be eafily heated without cracking, and is therefore of no great fervice, if hardened in the fire alone, and in its natural ftate: though this circumftance is of lefs inconveniency, than when it has original cracks, or is mixed with heterogeneous fubftances. The fteatites * is found purer and more folid in China than in any place in Europe. The natural faults of the European ones may, however, be altered by adding fome fat fubftance to it, when it is to be burnt; by which means it becomes black or brown; and this method is faid to be ufed at Bareith. The coarfe porcelane-like earth, which goes by the name of *French clay*, is ufed at the glafs-houfes, fteel furnaces, and other works of the fame nature, for the fame reafons as it is the

---

* The fteatites here meant is the fubftance of which the Chinefe joffes or figures commonly called rice figures are made: it is, according to my method, of the clafs of Talcs, and of the genus of Nephritics. D. C.

principal

principal ingredient in the making of crucibles, retorts, &c.

The boles have almoſt loſt their value as medicines, and are employed to make bricks, potters-ware, and pig-iron.

The tripoli is an indiſpenſible article for the poliſhing of metals, and ſome ſorts of ſtones; it is likewiſe on certain occaſions preferred for making moulds to caſt metals in.

The common clay is of the greateſt benefit in agriculture, except however the white clay and the fermenting clay, both of · Sect. xc. which varieties we know not yet how to apply to any uſe. By virtue of its coherency, this clay retains humidity, on which perhaps its chief benefit to vegetables depends, its other effects being occaſional, owing either to nature or art; unleſs the clay has formerly been a mould or *humus ater*, in which caſe it is juſt, that part of it ſhould enter again into the formation of the new vegetables. The clay uſed in the refining of ſugar, wants no other quality than that it may not dry too ſoon. But that ſpecies which is to be employed in fulling, muſt, if we were to judge *à priori*, beſides the fineneſs of its particles, be of a dry nature, or ſuch as attracts oils; though this quality may perhaps not be found in all thoſe clays which are now employed in that buſineſs.

S E C T. XCIII.

The FIFTH ORDER.

The Micaceous Kind, *Micaceæ*. The Glimmer, Daze, or Gliſt.

H 4   Theſe

Thefe are known by the following characters.

1. Their texture and compofition confift of thin flexible particles, divifible into plates or leaves, having a fhining furface.

2. Thefe leaves, or fcales, expofed to the fire, lofe their flexibility, and become brittle, and then feparate into thinner leaves : but in a quick and ftrong fire, they curl or crumple, which is a mark of fufion ; though it is very difficult to reduce them into a pure glafs by themfelves, or without addition.

3. They melt pretty eafily with borax, the microcofmic falt, and the alcaline falt ; and may, by means of the blow-pipe, be brought to a clear glafs, with the two former falts. The martial mica is, however, more fufible than the uncoloured ones.

There is not yet difcovered any loofe earth of this kind, but it is always found indurated.

SECT. XCIV.

A. Colourlefs or pure mica ; Daze, Glimmer, or Glift ; *Mica alba, five pura.*

1. Of large parallel plates, *Mica conftans lamellis magnis parallelis.* Mufcovy glafs, *Vitrum Mufcoviticum.*

Is tranfparent as glafs ; found in Siberia, and Eifdalen in the province of Wermeland.

2. Of fmall plates, *Mica fquamofa,* from Silfverberget, at Runneby, in the province of Blekinge.

3. Of

3. Of particles like chaff, or chaffy mica, *Particulis acerofis.*

4. Of twifted plates, crumpled mica, *Mica contorta, Talcum officinale.*

SECT. XCV.

B. Coloured and martial glimmer, *Mica colorata martialis.*

1. Of large parallel plates, *Mica lamellofa martialis.*

   a. Brown femi-tranfparent, from Kola, in Lapland.

2. Of fine and minute fcales.

   a. Brown.

   b. Deep green, from the mine of Salberg, in the province of Weftmanland.

   c. Light green, *Talcum officinale*, found in the ollaris, from Handol, in the province of Jemtland.

   d. Black, found in the granites, in the province of Upland.

3. Twifted or crumpled glimmer, *Mica contorta martialis.*

   a. Light green, in the ollaris, from Handol.

4. Chaffy glimmer, *Mica martialis particulis acerofis.*

   a. Black, is found in the ftone called *hornberg*, which occurs in moft of the Swedifh copper-mines; for inftance, thofe at Norberg, Flodberg, &c.

5. Criftallifed glimmer, *Mica drufica.*

   1. Of concentrated and erect fcales, *Drufa micacea conftans fquamis concentratis perpendicularibus caryophylloides.*

2. Of

2. Of hexagonal horizontal plates, *Drufa micacea conftans fquamis hexagonis horizontalibus.* This is found in the mines at Salberg, in the province of Weſt-manland.

SECT. XCVI.

OBSERVATION on the MICÆ, or GLIMMERS.

The ftones belonging to this order are by moſt authors confidered as Apyri, which they really are in fome degrees of heat, and when they are mixed with certain bodies : but they may at the fame time with equal propriety be called Vitre-fcents, both *per fe* or by themfelves, becaufe they melt with that degree of fire in which neither quartz nor limeftone are in the leaſt altered; and are ſtill more readily fufed, when mixed with a martial earth, either by nature or art: hence, if *hornberg* is naturally mixed with copper ores, as is frequently found in Sweden, it is no way detrimental to the fmelling of them, as they commonly contain a fufficient quantity of fulphureous acid, which fcorifies the iron. But when the glimmer is mixed with quartz, it may perhaps be impof-fible to melt it, becaufe it renders the quartz fo compact, as to prevent it from cracking, which may be feen on the rock-ftone (Sect. cclxii.): The mica does the fame, when it is interfperfed in an apyrus clay ; and this is the reafon why the ollaris fo ſtrongly refifts the fire.

The mica has in fome degree the fame qualities as an argillaceous earth ; but, for want of fufficient experiments and obfervations, we cannot yet affert it to be a product of clay.

The

The martial mica in a calcining heat acquires a yellow shining colour, which has induced many to examine it for gold; but nothing can be obtained from it except iron, which may be diffolved or extracted by means of aqua regis: although a late German author has pretended that he produced from the mica an unknown femi-metal, which refembled iron mixed with zink. Nevertheless he owns, that he has not examined this femi-metal, and that for obtaining it he ufed a flux, compofed of feveral metals, fome of which probably united with the iron in the mica: wherefore it is probable we fhall never hear more of it.

Some of the micaceous kind feem fat and unctuous, and others harfh and dry: it is not improbable that the former may contain a phlogifton, although this cannot be extracted from them in form of a pure *oleum talci*; in other particulars, they are fo like one another, that there is no reafon for making them two diftinct genera.

The *talc cubes*, as they are called, which have the figure of alum, and are fometimes found in the copper-mine of Falun, in the province of Dalarne, and which are very much valued by fome foffilogifts, are, when broke, found to confift of an iron ore, often mixed with a yellow or marchafitical copper ore, and only covered with a very thin coat of mica.

The tranfparent Mufcovy glafs is ufed for windows, and upon all occafions where panes of glafs are wanted. Perhaps it might alfo be advantageoufly employed to cover houfes.

The twifted or crumpled mica, which is found at Handol in Jemtland, is there manufactured into kettles and other veffels, as-alfo for hearths of chimnies; and the powder which falls in the working

working of this ftone may be mixed with the common falt, for the diftillation of the muriatic acid.

SECT. XCVII.

The Sixth Order.

The Fluors *, *Fluores Minerales*. Suet. *Fluff-arter*. Germ. *Fluff-arten*.

Thefe are commonly called fluxing vitref-cent, or glafs fpars, becaufe moft part of them have a fparry form and appearance : they are, however, often met with in an indeterminate figure.

Thefe are only known in an indurated ftate, and diftinguifh themfelves from the other earths, by the following characters.

1. They are fcarce harder than a calcareous fpar, and confequently do not ftrike fire with the fteel.

2. They do not ferment with acids, neither before nor after calcination, notwithftand-ing a phlogifton or an alcali had been added in the calcination.

3. They do not melt by themfelves, but only fplit to pieces when expofed to a ftrong fire †. But,

4. In mixtures with all other earths, they are very fufible, and efpecially when they are

---

* I have adopted the name of Fluors, in Englifh, to this order. D. C.

† There may, perhaps, be fome fluors that are pretty re-fractory in the fire, fo as not to be melted : however, all thofe which I have tried, have melted pretty eafily by the blow-pipe ; but I have always taken great care in thefe ex-periments, that they might not fly away before they were heated through. E.

blended

blended with the calcareous earth, with which they melt to a corroding glafs, which diffolves the ftrongeft crucibles, unlefs fome quartz or apyrus clay is added thereto.

5. When heated flowly, and by degrees, they give a phofphorefcent light : but as foon as they are made red-hot, they lofe this quality. The coloured ones, and efpecially the green, give the ftrongeft light, but none of them any longer than whilft they are well warm.

6. They melt and diffolve very eafily by the addition of borax, and next to that by the microcofmic falt, without ebullition.

SECT. XCVIII.

A. Indurated Fluor, *Fluor mineralis induratus.*

1. Solid, of an indeterminate figure, *Fluor particulis impalpabilibus, figurâ indeterminatâ.*

Is of a dull texture, femi-tranfparent, and full of cracks in the rock.

a. White, found in Batgrufvan, at Yxfio in Nya Kopparberget in Weftmanland.

SECT. XCIX.

2. Sparry Fluor, *Fluor Spatofus.*

It has nearly the figure of fpar, though, on clofe obfervation, it is found not to be fo regular, nothing but the glofly furfaces of this ftone giving it the refemblance of fpar.

a. White,

a. White, found in Stripas at Norberg, in the province of Weftmanland.

b. Blue, from Norrgrufve, at Wefterfilf-verberget in Weftmanland.

c. Violet, from Diupgrufvan, at the laft mentioned place, and alfo from Stripas and Fogerlid; and Giflof in the province of Skone.

d. Deep green, from Stollberget in Stora Kopparbergflan.

e. Pale green, from Kuppgrufven, at Garpenberg in the province of Dalarne.

f. Yellow, from Giflof in Skone.

SECT. C.

3. Cryftallifed Fluor, *Fluor cryftallifatus*, when in fingle criftals; but Fluor Drufe, when many criftals are heaped together.

1. Of an irregular figure.
   a. White.
   b. Blue, both from Norberget and Norberg in Weftmanland.
   c. Red, from Heflekulla iron-mine, in the province of Nerike.

2. Of a cubical figure.
   a. Yellow, and
   b. Violet, from Giflof in Skone, Blyhall in the parifh of Barkaro in the province of Weftmanland.

3. Of a polygonal fpherical figure.
   a. White, from Bockbackeveggen in Falun copper-mine in Dalarne.
   b. Blue, from Bondgrufvan, at Norberg in Weftmanland.

4. Of

4. Of an octoëdral figure.

 a. Clear, colourless. This I have feen in the collection of the mine-mafter Mr. Von Swab.

SECT. CI.

OBSERVATION on the FLUORS.

There are not yet any probable reafons given, why thefe ftones fhould be ranked amongft the calcareous or any other earths; and if I am not quite miftaken in my judgment, they are fo much the more different from the calcareous earth, as they, when melted together with it, produce an effect which never can be afcribed to the alcaline earths; not to mention, that there is by no method yet known any calcareous fubftance to be extracted from them, nor is there any poffibility of decompounding them.

That which caufes the phofphorefcent light vanifhes in the fire, it being impoffible to collect it: in the prefent ignorance of the nature of this matter, it cannot therefore be afferted, whether it is one of the conftituent parts neceffary to the compofition of thefe ftones, or if, in regard to its fmall quantity, it even deferves any attention. I take it to be a fubtle phlogifton, which being modified in various manners, gives rife to fuch various colours.

At mineral works this kind of ftone is very ufeful in promoting the fufion of the ores, and is therefore as much valued by the fmelters, as the borax is by the effayers: it has alfo from this quality got the name of *fluor*, or *flux*.

The refemblance between the coloured fluors, and the compofitions made of glafs, has perhaps contributed

contributed not only to the fluors being reckoned of the fame value as the coloured quartz cryftals, by fuch collectors as only mind colour and figure; but alfo to their obtaining a rank among the precious ftones in the apothecaries and druggifts fhops. They, however, may be permitted to enjoy that honour, fince our modern phyficians do not make more ufe of them than of the others.

SECT. CII.

The SEVENTH ORDER.

The Afbeftus Kind, *Afbeftinæ.*
Thefe are only yet difcovered in an indurated ftate : their characters are as follow.

1. When pure, they are very refractory in the fire.
2. In large pieces they are flexible.
3. They have dull or uneven furfaces.
4. In the fire they become more brittle.
5. They do not ftrike fire with the fteel.
6. They are not attacked by acids.
7. They are eafily brought into fufion by borax.

In this order are included both thofe varieties which by foffilogifts have been mentioned under the names of *Amianti* and *Afbefti,* and have often been confounded together.

SECT.

SECT. CIII.

1. Afbeftus which is compounded of foft and thin membranes, *Afbeftus membranaceus ;* *Amiantus Wallerii.*

A. Of parallel membranes, *Afbeftus membranis conftans parallelis : Corium,* five *Caro Montana,* Mountain-leather.
1. Pure.
   a. White, from Salberg in Weftmanland.
2. Martial.
   a. Yellowifh brown, from Storrginningen, at Dannemora, in the province of Upland.
   This melts pretty eafily in the fire to a black flag, or glafs.

SECT. CIV.

B. Of twifted foft membranes, *Afbeftus membranis conftans contortis : Suber montanum,* Mountain cork.
1. Pure.
   a. White, from Salberg in Weftmanland.
2. Martial.
   a. Yellowifh brown, from Dannemora.
   This has the fame quality in the fire as the martial mountain leather.

SECT. CV.

2. Of fine and flexible fibres, *Afbeftus fibrofus : Afbeftus,* or Earth Flax, *Afbeftus Wallerii.*

I   *A.* With

A. With parallel fibres, *Asbestus fibris conftans parallelis: Byssus.*

1. Pure and foft.

    a. Light green, from Schelkowa Gora in Siberia.

    b. White, from Ulrica's Ort in the mine of Salberg in Weftmanland: it is there found together with mountain leather.

2. A little martial, and more brittle.

    a. Greenifh, from Baftnas Grufva, at Ryddarhyttan in Weftmanland. There it forms the greateft part of the vein out of which the copper ore is dug ; a great part of it is confequently melted together with the ore, and is then brought to a pure femi-tranfparent martial flag or glafs.

SECT. CVI.

B. Of broken and recombined fibres, *Asbeftus fibris conftans abruptis et conglutinatis.*

1. Martial.

    a. Light green, from Baftnas Grufva at Riddarhyttan.

SECT. CVII.

It has been already obferved under the title of Cockle, or Shirl, (Sect. lxxiv.) that the afbeftus is often confounded with it.

OBSERVATION on the ASBESTUS KIND.

I am much inclined to believe that the Afbefti, as well as the Micæ, are produced from an argillaceous

gillaceous earth, both becaufe they become brittle in the fire, which is a proof that they harden, and becaufe they become more fufible by the admixtion of a martial earth: but the method nature makes ufe of for this change is as unknown, as it might perhaps in other refpects be neceffary, hot to force the earths together, for fome flight reafon, within the compafs of a few orders.

The Siberian Afbeftus, which may be confidered as the principal and chief of the fibrous kind, is, as it were, confumed by the flame of a blowpipe, and does not leave any more certain mark of fufion; but it melts readily with borax to a clear and colourlefs glafs.

The natural ftore of this kind is in proportion to its œconomical ufe, both being very inconfiderable. It is an old tradition, that in former ages they made cloaths of the fibrous afbefti, which is faid to be expreffed by the word *Byffus*; but it is not very probable, fince, if one may conclude from fome trifles now-a-days made of it, as bags, ribbons, and other things, fuch a drefs could neither have an agreeable appearance, nor be of any conveniency or advantage. It is more probable that the Scythians dreffed their dead bodies, which were to be burnt, in a cloth manufactured of this ftone; and this has perhaps occafioned the above fable.

Paper is likewife made from this ftone, only to fhew its fixity in the fire, and to procure fome efteem and value to this curious fubftance.

It was reported fome years ago, that the French fearched for afbefti, in order to mix it with the tar for preferving houfes and fhips; but the queftion is, If the afbefti can be of more fervice than pounded mica, or charcoal-duft employed to the fame purpofe?

I 2 SECT.

SECT. CVIII.

The EIGHTH ORDER.

Zeolites.

This is defcribed in its indurated ftate, in the Tranfactions of the Academy of Sciences at Stockholm, for the year 1756, and there methodifed as a ftone *fui generis*, in regard to the following qualities.

1. It is a little harder than the fluors, and the calcareous kind: it receives however fcratches from the fteel, but does not ftrike fire with it.

2. It melts eafily by itfelf in the fire, with a like ebullition as borax does, into a white frothy flag, which not without great difficulty can be brought to a folidity and tranfparency.

3. It is eafier diffolved in the fire by the mineral alcali *(fal fodæ)*, than by the borax and microcofmic falt.

4. It does not ferment with this laft falt, as the lime does; nor with the borax, as thofe of the gypfeous kind.

5. It diffolves very flowly, and without any effervefcence, in acids, as in oil of vitriol and fpirit of nitre. If concentrated oil of vitriol is poured on pounded zeolites, a heat arifes, and the powder unites into a mafs *.

* Since the publication of this Effay, there has been difcovered more varieties of the zeolites, particularly at Adelfors's gold-mines in Smoland in Sweden, of which fome forts do not melt by themfelves in the fire, but diffolve readily in the acid of nitre, and are turned by it into a firm jelly. E.

6. In

6. In the very moment of fufion it gives a phofphorus or light.

SECT. CIX.

The zeolites is found in an indurated ftate.
1. Solid, or of no vifible particles, *Zeolites folidus particulis impalpabilibus.*
A. Pure, *Zeolites durus.*
    a. White, from Iceland.
B. Mixed with filver and iron.
    a. Blue, *Lapis lazuli*, from the Buckarian Calmucks.

This, by experiments made with it, has difcovered the following properties.

1. It retains for a long time its blue in a calcining heat, but is at laft changed into a brown colour.
2. It melts eafily in the fire to a white frothy flag ; which, when expofed to the flame of a blow-pipe, is greatly puffed up, but in a covered veffel, and with a ftronger heat, becomes clear and folid, with blue clouds in it.
3. It does not ferment with acids : but,
4. Boiled in the oil of vitriol, it diffolves flowly, and lofes its blue colour.

When a fixed alcali is added to this folution, a white earth is precipitated, which being fcorified with borax, yields a filver regulus, that varies in bignefs, according to the various famples of the ftone.

I 3   5. By

5. By ſcorification with lead, there has been extracted two ounces of ſilver out of a hundred pounds weight of the ſtone.

6. The preſence of ſilver is not diſcovered with the ſame certainty by the ſpirit of nitre as by the oil of vitriol.

7. When the ſpirit of ſal ammoniac is added to any ſolution, made either of crude, or of a perfectly calcined lapis lazuli, there is no blue colour produced ; which proves that this colour is not owing to copper, as ſome have pretended : and this is farther confirmed by the fixity of the blue colour in the fire (1, 2.), and by the colour of the ſlag or glaſs (2.).

8. It is a little harder than the other kinds of zeolites, but does not however in hardneſs approach to the quartz, or to other ſtones of the ſiliceous kind in general ; becauſe the pureſt and fineſt blue lapis lazuli may be rubbed with the ſteel to a white powder, although it takes a poliſh like marble.

9. The lapis lazuli, when perfectly calcined, is a little attracted by the loadſtone ; and ſcorified with lead, the ſlag becomes of a greeniſh colour, not ſuch a colour as copper gives, but ſuch as is always produced by iron mixed with a calcareous ſubſtance *.

---

* The lapis lazuli is ſeldom found pure, but is moſt generally full of veins of quartz, limeſtone, and marcaſite : however,

SECT. CX.

2. Sparry Zeolites, *Zeolites fpatofus.*
This refembles a calcareous fpar, though it
is of a more irregular figure, and is more
brittle.

a. Light red, or orange-coloured, from Nya
Krongrufvan, one of the gold-mines at
Adelfors, in the province of Smoland.

ever, for thefe experiments none but the pureft pieces have
been picked, fuch as have been examined through a magnifying glafs, and been judged as free from heterogeneous
mixtures as poffible. It is to be wifhed, that thofe who have
a fufficient quantity of this ftone would continue thefe experiments, in order to difcover what fubftance it is that makes
this blue colour, which is fo conftant in the fire, fince it cannot depend either on copper or iron ; for though thofe metals,
on certain occafions, give a blue colour, yet they never produce any other but what inftantly vanifhes in the fire, and is
deftroyed by means of an alcali. What is mentioned in feveral books about the preparation of the ultramarine from
filver, can by no means be objected here, fince in thofe proceffes the filver employed is mixed with copper, and other
fubftances, which contain a volatile alcali, whereby the blue
colour is produced ‡.
In regard to the above-mentioned qualities of this ftone, it
cannot be claffed under any other kind of earth than this.

‡ Mr. Margraf has fince, in his Chemical Differtations, printed in
German in the year 1761, publifhed fome experiments on the lapis lazuli ;
and in the chief agrees with our author, without, however, knowing any
thing of thefe his experiments. Mr. Margraf alfo proves that there is no
copper in this ftone ; and befides tells us, that he has found both a calcareous and a gypfeous fubftance in it, although he took care to pick out
only the very pure bits for his experiments However, I am led to imagine,
that the calcareous fubftance is not effential to the exiftence of the lapis
lazuli, fince Mr. Cronftedt exprefly fays, that the ftone he tried did not
ferment at all with acids. He farther mentions this remarkable circumftance, which makes it ftill more evident that the lapis lazuli belongs to
the zeolites, viz. that, when calcined and diffolved in the acids of vitriol,
of common falt, and of nitre, it turned all thofe acids into a jelly.
However, he does not take any notice of its containing any filver, tecaufe he
did not profecute his experiments fo far on that point ; but fome of his
experiments, neverthelefs, feem to indicate, as if all forts of lapis lazuli did
not contain filver. E.

SECT. CXI.

3. Criftallifed Zeolites, *Zeolites cryflallifatus.*
Is more common than the two preceding kinds, and is found,

A. In groupes of criftals in form of balls, and with concentrical points, *Cryftalli zeolitis pyramidales concreti ad centrum ten-- dentes.*

  a. Yellow, from Swappawari, in Tornea in Lapland.

  b. White, from Guftavfgrufvan, in the province of Jemtland.

B. Prifmatical and truncated criftals, *Cryftalli zeolitis diftincti figura prifmatica truncata.*

  a. White, from Guftavfgrufvan in Jemtland.

C. Capillary criftals, *Cryftalli zeolitis capillares.*

  Are partly united in groupes, and partly feparate. In this latter accretion they refemble the capillary, or feather filver ore (Sect. clxxiii.), and is, perhaps, fometimes called *Flos ferri,* at places where the nature of that kind of ftone is not yet fully known.

  Thefe criftals are found,

  a. White, from Guftavfgrufvan in Jemtland.

SECT. CXII.

Observation on the Zeolites.

This kind of ftone has nearly the fame qualities in the fire as the boles (Sect. lxxxv) ; fo that both

of

of them, when more nicely examined, may perhaps be found to belong to the fame order, and perhaps be fome kind of earth, whofe properties have been long and perfectly known.

The *terra porcellanea Luneburgica*, which Bruckman mentions, and Mr. Wallerius has ranked among the gypfa, may, perhaps, belong to this order : but I have not been able to procure a fpecimen of it, to compare it with the zeolites, which alfo is very fcarce, not being found in our country except in very fmall veins and cavities. To this fcarcity is owing, that it has not yet been tried in the fire together with other kinds, except with the fparry fluor. With that it does not fufe very readily, becaufe, when equal parts of them are melted together, an opaque flag or glafs is produced of the fame colour with the alcali of nitre, of a fibrous texture, and of an uneven furface.

The quality of fwelling in the fire, like the borax, is peculiar to the criftals, (Sect. cxi.) becaufe the other varieties rife only into fome fmall blifters, which are of a white colour at their edges, and inftantly cover themfelves with a white glaffy fkin, after which they become quite refractory.

SECT. CXIII.

The NINTH ORDER.

The Manganefe Kind, *Magnefia*.

The ftones belonging to this order, are in Swedifh called *Brunften*, in Latin *Syderea*, or *Magnefia nigra*, in order to diftinguifh them from the *Magnefia alba officinalis*, and in French *Mangonefe*, &c. They are by fome

litho-

lithographifts entirely omitted, and by others ranked among the iron ores; but, as I am convinced both by my own experience, and by that of others, that they contain no greater quantity of metal than fometimes two or three per cent. of iron, and fometimes a little tin, I think that the remaining part, which muft confequently be confidered as a kind of earth, deferves its particular feparate place in a mineral fyftem, at leaft until a farther infight into its nature may be obtained: and to this opinion I have been perfuaded by its following peculiar qualities:

1. The manganefes confift of a fubftance, which gives a colour both to flags, and to the folutions of falts, or, which is the fame thing, both to dry and to liquid menftrua; viz.

   a. Borax, which has diffolved manganefe in the fire, becomes tranfparent, of a reddifh brown or jacinth colour.

   b. The microcofmic falt becomes tranfparent with it, of a crimfon colour, and moulders in the air.

   c. With the fixed alcali, in compofitions of glafs, it becomes violet; but if a great quantity of manganefe is added, the glafs is in thick lumps, and looks black.

   d. Scorified with lead, the glafs gets a reddifh brown colour.

   e. The lixivium of a deflagrated manganefe is of a deep red colour.

2. It deflagrates with nitre, which is a proof that it contains fome phlogifton.

3. When reckoned to be light, it weighs as much as an iron ore of the fame texture.

4. Being melted together with glafs compofitions, it ferments during the folution: but

it

it ferments in a ftill greater degree, when it is melted with the microcofmic falt.

5. It does not excite any effervefcence with the fpirit of nitre: aqua regia, however, extracts the colour out of the black, and diffolves likewife a great deal of it, which, by means of an alcali, is precipitated to a white powder.

6. Such colours as are communicated to glaffes by manganefe, are eafily deftroyed by the calx of arfenic or tin: they alfo vanifh of themfelves in the fire.

7. It is commonly of a loofe texture, fo as to colour the fingers like foot, although it is of a metallic appearance when broke.

SECT. CXIV.

Manganefe is found,

A. Loofe and friable, *Magnefia friabilis terriformis.*

a. Black, feems to be weathered or decayed particles of the indurated kind, from England.

SECT. CXV.

B. Indurated, *Magnefia indurata.*

1. Pure, in form of balls, whofe texture confifts of concentric fibres, *Magnefia pura fphærica radiis concentratis.*

a. White, *Magnefia alba ftri&è fic dicta,* is very fcarce. I have feen a fpecimen of this kind in a collection from an unknown place in Norway; and by examining a piece of it, I found that it differed from
the

the common manganefe, by giving to
the borax a deep red colour in the fire :
this fort acquires a reddifh brown colour when it is calcined.

b. Red manganefe is faid to be found in
Piedmont. This I have never feen ; but
I have been told by an ingenious gentie-
man, that this variety is free from iron,
and gives to glafs rather a red than a violet colour.

S E C T. CXVI.

2. Mixed with a fmall quantity of iron,
Magnefia parum martialis.

a. Black manganefe, with a metallic brightnefs. This is the moft common kind,
and is employed at the glafs-houfes, and
by the potters.

It is found,

1. Solid, of a flaggy texture, *Magnefia
textura vitrea*, from Skidberget, in
the parifh of Lekfand, in the province
of Dalarne.

2. Steel grained, alfo from Skidberget.

3. Radiated, *Radiata*, ftill from Skidberget, and Tiveden, in the province
of Oftergottland.

4. Criftallifed.

a. In form of coherent hemifpheres,
Hemifpheriis continuis, from Skidberget in Lekfand.

S E C T. CXVII.

3. Blended with a fmall quantity of iron
and tin, *Magnefia parva cum portione martis*

tis et jovis mixta : Spuma Lupi, or *Wolfram* *.

1. With coarse fibres.

   a. Of an iron colour, from Altenberg in Saxony. This gives to the glass compositions, and also to borax and the microcosmic salt, an opaque whitish yellow colour, which at last vanishes.

SECT. CXVIII.

OBSERVATIONS on the MANGANESE.

Though it may seem difficult to many, to distinguish the kinds of manganese by their appearance, or external marks; yet it is extremely easy to know them by experiments made in the fire, if attention is had to the above-mentioned phenomena (Sect. cxiii.). From hence it is not difficult to comprehend why manganese has hitherto been either omitted, or erroneously ranked in systems, viz. because it has, like many other mineral bodies, been examined only by sight, while the more troublesome method of examining it in the fire, has been overlooked.

Some might perhaps imagine the manganese to be the remainder of some metal, which cannot be reduced again into its metallic state; but it ought to be remembered, that no metal can, by any means yet known, be brought to an absolutely irreducible earth or calx, unless perhaps by the burning-glass, and therefore there is no reason to suspect that nature gives such a production. Ig-

---

* Wolfram is a name which is also sometimes given to mock lead, and sometimes to cockle, or shirl, as also to other minerals; however, it is chiefly given to this species of manganese, when it occurs in the tin-mines. E. and D. C.

norance

norance and idlenefs have invented certain terms
or expreffions, to avoid giving an account of thofe
ores or mineralifations, which are not eafy enough
to be decompounded ; for inftance, *wild, rapacious, arfenical, volatile,* &c. and fome iron ores
in particular have been thus called ; by which
means it has happened, that œconomical reflections have often been added to natural and philofophical defcriptions : and thus others are, de-
terred from examining many bodies, of which we
have got, and ftill retain falfe notions by this way
of proceeding.

The manganefe has by fyftematifts been commonly ranked among fuch iron ores ; but the
artificers who make ufe of it in the manufacture
of glafs do not know it ; nor can they by any
means be perfuaded to ufe any of the pretended
bodies a-kin to it, inftead of the manganefe itfelf,
fince experience prevails more with them than
fuppofitions. The confumption of the manganefe
is but fmall, and therefore it is not a very profitable article.

SECT. CXIX.

The SECOND CLASS.

The SALTS, *Salia.*

By this name thofe mineral bodies are called,
which can be diffolved in water, and give it a
tafte ; and which have the power, at leaft when
they are mixed with one another, to form new
bodies of a folid and angular fhape, when the
water in which they are diffolved is diminifhed
to a lefs quantity than is required to keep
them

them in folution; which quality is called Criftal-lifation *.

SECT. CXX.

In regard to the known principal circumftances or qualities of the mineral falts, they are divided into

1. Acid Salts, or Mineral Acids, *Salia Acida.*
2. Alcaline Salts, or Mineral Alcalis, *Salia Alcalina.*

The FIRST ORDER.,

Acid Salts, *Salia Acida.*

The characters of thefe falts are, that they,

1. Have a four tafte.
2. Are corrofive; that is to fay, have a power of diffolving a great number of bodies.

* No other falts ought to be confidered and ranked in a mineral fyftem, but thofe which are found natural in the earth (Sect. i.) ; and for this reafon a great number of falts will be in vain looked for here, viz. all fuch as are either natural or prepared by art in the other two kingdoms of nature, and from fubftances belonging to them. Amongft thefe is nitre itfelf, and its acid, and the vegetable acid, fince thefe are never had from true mineral bodies ; nor is it demonftrated, that they have their origin from the true mineral vitriolic and muriatic acids. There have, indeed, been many attempts made to reduce moft of them to a vitriolic acid, which by many is called the univerfal acid : but experiments will not agree with it ; at leaft nobody has yet been able, by uniting a phlogifton with another acid than the true vitriolic, to produce any fubftance in every particular refembling the true brimftone, or fulphur. For this reafon I cannot yet give my affent to Doctor Pietfch's opinion, who endeavours to prove, that the acid of nitre is derived from the vitriolic acid, that is, before his theory is confirmed by experience in the large way, and the analyfis has been more plainly laid open : but I think

the

3. They have a ſtrong attraction to the
alcaline ſalts and earths, whence they always unite with them with an effervefcence, and ſometimes with a ſtrong heat:
by this mixture bodies are produced,
which are employed in common life under
the names of *vitriols, neutral ſalts, gypſum,* &c.

4. They change moſt of the expreſſed blue
juices of vegetables into red.

the queſtion remains ſtill undecided, if the nitrous, vegetable,
and urinous acids are primitive ſubſtances? or if they owe
their origin to one and the ſame principle? and, if this laſt
be the caſe, of what nature this principle is? But howſoever
this may prove, the conſideration of theſe acids ſeems more
properly to belong to another ſcience. The ſame may be
ſaid of the doctrine which holds, that the nitre is produced
from the principles of the ſea-ſalt, by a certain peculiar modification.

The above-mentioned two mineral acids, whoſe qualities
we know nothing of, until they have been by art extracted
from the vitriols, and the ſea-ſalt, are indeed never found
pure in nature, becauſe as ſoon as they, on any occaſion, are,
either by a natural or artificial heat, ſeparated from any ſubſtance, they inſtantly attack and unite with another. Ne-
vertheleſs, as they may, and perhaps ſometimes really do
exiſt in form of vapours, which eſcape our ſight; and that
the theory of the ſalts, and the ſaline ores, is founded upon
qualities already diſcovered in theſe acids; I have thought it
neceſſary to deſcribe them ſuch as they are, when mixed with
pure water alone; and this the rather, ſince the water is their
moſt common vehicle, in the exerciſe of their effects in the
mineral kingdom.

It has been obſerved before (Sect. xi.), that the qualities of
arſenic in form of a calx may agree with the definition of
the ſalts, and at the ſame time be reckoned among the ſemi-
metals, which cannot be any otherwiſe explained, than that
the arſenic conſidered in a certain reſpect and form, is a ſalt;
and when conſidered in other circumſtances, a metal. This
is the caſe with ſeveral other bodies of the mineral kingdom.

5. They

5. They separate the alcali from the fat, when they have been united in foap; which effect is called to *turdle*, or coagulate.

6. They are volatile and fubtile, fo as never to be obfervable by the naked eye, unlefs they are mixed with heterogeneous bodies; and therefore the figure of the pure mineral acids cannot be defined but by guefs.

SECT. CXXI.

A. The vitriolic acid, *Acidum vitrioli aluminis et fulphuris.*

I. The pure vitriolic acid, *Acidum vitrioli purum.*

Is, in abftract, confidered as poffible to occur in nature: its qualities, when mixed with water, in which it is caught by diftillation, are as follows.

1. When mixed with the leaft poffible quantity of water, it is of an unctuous appearance, and is for that reafon improperly called oil of vitriol.

2. It has in that ftate a confiderable heavinefs, viz. in comparifon to water, as 1700 to 1000.

3. It diffolves filver, tin, the regulus of antimony, and quickfilver; but,

4. When mixed with more water, it diffolves zinc, iron, and copper.

5. It diffolves likewife the calcareous earth, and precipitates with it in form of a gypfum, of which a part fhoots into gypfeous Drufen, *Selenites et cryftalli gypfei.*

6. It unites with the earth of quartz, when it has been previoufly diffolved in the

K *liquor*

liquor filicum; and with a pure argillaceous earth, diffolving it without any
fermentation: with both thefe earths it
makes alum.

7. It has a ftronger attraction to the inflammable fubftance, than to the alcaline
falt, and forms with it a body, which
properly may be called the *mineral*
fulphur.

8. When it is perfectly united with phlogiftic fubftances belonging to the vegetable kingdom, and the water has
been perfectly feparated, this mixture
catches flame in the open air, and is
confumed, as may be feen by the powder
called *Pulvis pyrophorus*.

9. It attracts water ftrongly, and the
aqueous vapours out of the air: and if
a great quantity of water is added to it
at once, a ftrong heat arifes.

10. It unites readily and eafily with the
alcalis, whereby, according to their nature, different compounds are produced,
which have obtained the names of *Tar-*
tarus vitriolatus, fal mirabile, and *fal am-*
moniacum fixum.

SECT. CXXII.

. The vitriolic acid mixed or faturated,
Acidum vitrioli aliis corporibus faturatum.
A. With metals, *Metallis faturatum. Vitri-*
ola, Vitriols.
   a. Simple vitriols, *Vitriola fimplicia.*
     1. Martial vitriol, green vitriol or cop-
     peras, *Vitriolum martis fimplex.*
                     This

This is the common green vitriol, which naturally is found diffolved in water, and is produced in abundance by decayed or calcined marcafites.

2. Copper vitriol, blue vitriol, *Vitriolum Veneris feu Cypricum.*

This is of a deep blue colour, and is found in all Ziment waters, as they are called; for inftance, at Neufohl in Hungary, in St. Johan's mine at Fahlun in the province of Dalarne, at Nya Kopparberget in Weftmanland, and the coppermines at Wicklow in Ireland, &c. It is however feldom perfectly free from an admixture of iron and zink.

3. Zink vitriol, *Vitriolum zinci.*

Is white and clear as alum, and is found at the Rammelfberg in the Hartz, as alfo in the rubbifh at Stollgrufvan in Weftmanland, where the mock lead has decayed either fpontaneoufly, or after having been burnt.

SECT. CXXIII.

b. Compound vitriols, *Vitriola compofita.*

1. Vitriol of iron and copper, *Vitriolum ferrum et cuprum continens.*

Is of a bluifh green colour.

2. Vitriol of iron, zinc, and copper, *Vitriolum ferrum zincum et cuprum continens.*

This

This verges more to the blue than to the green colour. It is made at Fahlun in Dalarne, from the water which is pumped out of the copper-mines: in this water large criftals of vitriol are often ready formed. If this vitriol is dipped in water, and afterwards rubbed on clean iron, the copper does not precipitate from it.

3. Vitriol of zinc and iron, *Vitriolum zinco-ferreum.* This is the green vitriol from Goflar in the Hartz.

4. Vitriol of zinc and copper, *Vitriolum cupreo zinceum.* This is the blue vitriol from Goflar.

5. Vitriol of nickel and iron, *Vitriolum ferrum et niccolum continens.*

Is of a deep green colour, and is contained in the ochre or decayed parts of the Nickell, at the Cobalt-mines at Los, in the province of Helfingland*.

* Moft part of the vitriols owe their formation to art : be-caufe when fuch ores as contain fulphur, are dug out of the mines by means of fire, the phlogifton of the fulphur is by the heat expelled, leaving the acid behind, which, being let loofe or freed, is thereby enabled to attract and unite with watry vapours, diffolving at the fame time the metals; and it is thus the vitriols are formed. Every fort of ore does not commonly decay or weather in a natural manner, without being promoted by art ; and this decaying or weathering is moftly performed in the open air ; for which reafon no very great quantity of vitriol can be expected in that way: for when any ore thus weathers or decays, the diffolved particles are by degrees carried off by the rain, and are at laft found in a diffolved ftate in certain fprings or mineral waters. All fuch ores may therefore be called true vitriol ores, as contain iron, copper, zinc, and nickel mineralifed with fulphur. The acid in the vitriols, however, is not dulcified by the metals, as it is by the alcali in the true neutral falts.

SECT.

SECT. CXXIV.

B. The acid of vitriol mixed or faturated with earths, *Acidum vitrioli terris mixtum feu faturatum.*

1. With a calcareous earth. Gypfum. See Sect. xiii.
2. With an argillaceous earth. The Alum kind, *Alumina.*

 a. With a fmall quantity of clay, *Acidum vitrioli argillâ faturatum.* Native or plumofe alum, *Alumen nativum five plumofum.*

 Is found on decayed alum ores in very fmall quantities ; and therefore through ignorance the alabaftrites and felenites, both of which are found among moft of the alum flates, are often fubftituted in its ftead ; as is alfo fometimes the afbeftus, notwithftanding the great difference there is between the alum and thefe, both in regard to their ufes and effects *.

 b. With a greater quantity of pure clay, *Argilla pura acido vitrioli imbuta.* White alum ore, *Minera aluminis alba.*

---

* The gypfa and afbefti, but more efpecially the latter, have been ufed through ignorance, in moft countries, for plumofe native alum ; and the fort fold formerly in the fhops for it was a greenifh white kind, from Germany, very rigid, but extremely brittle, and breaking into fpicula or prickles. Selenites was never fubftituted for alum ; and the reafon the afbefti and fibrofe gypfa were fubftituted for it, was only on account of the fimilarity of ftructure, not, as our author fays, on account of their being found together. See my Lectures. D. C.

1. Indu-

1. Indurated pale red alum ore, *Schiſtus aluminis Romanus*. Is employed at Lumini, not far from Civita Vecchia in Italy, to make the pale red alum called Roach Alum. This is, of all alum ores, the moſt free from iron; and the reddiſh earth which can be precipitated from it, does not ſhew the leaſt marks of any metallic ſubſtance.

c. With a very large quantity of martial clay, which likewiſe contains an inflammable ſubſtance, *Argilla martialis et phlogiſtica acido vitrioli imbuta.* Common alum ore.

Is commonly indurated and flaty, and is therefore generally called Alum Slate, *Schiſtus aluminoſus ater et bruneſcens.*

It is found,

1. Of parallel plates, with a dull ſurface, *Schiſtus lamelloſus regularis*, from Andrarum in the province of Skone, Hunneberg and Billingen in the province of Weſtergottland, Rodoen in the province of Jemtland, and the iſland of Oeland, &c. *

2. Undulated and wedge-like, with a ſhining ſurface, *Schiſtus aluminoſus undulatus et cuneiformis fiſſuris ſplendentibus.*

This at the firſt ſight reſembles pit coal; it is found in great abundance in the pariſh of Nas in Jemtland †.

___

* In England, the great alum works at Whitby, in Yorkſhire, are of this kind. D. C.

† The purity above-mentioned (*b.* 1.) of the earth of the Roman or roach alum, is meant with the ſame reſtriction

a5

SECT. CXXV.

C. Vitriolic acid united with phlogiston, *Acidum vitrioli phlogisto combinatum.* The sulphur kind, *Sulphura.* See Sect. cli.

SECT. CXXVI.

D. Vitriolic acid saturated with alcaline salt, *Acidum vitrioli alcali minerali saturatum.*

as in general is understood by that expression, viz. that the heterogeneous particles are not very obvious, nor of any great consequence.

The phlogiston which is contained in the black alum slates, may perhaps during the calcination dispose the iron to be easier dissolved ; and it may also occasion the black colour in some of them, that even contain but very little of iron, as most likely in part of those from Nas (*c.* 2.).

It is not easily determined, whether the earth in the alum slates is argillaceous or quartzose, or whether it is a black indurated *humus,* or mould, because all those three earths, when dissolved in the vitriolic acid, produce alum. The Cologne pipe-clay is a plain proof of the first ; the quartzose earth, in *liquor silicum,* of the second ; and aluminous fossil-woods are actually employed for making alum in Bohemia and Hesse. These earths may, or may not, contain iron ; however, they prevent, in the former case, the phlogiston, together with the vitriolic acid, from mineralizing all the iron, and making a marcasite of it, excepting here and there, in some insignificant quantity, as in cracks, or when it meets with some heterogeneous bodies, as shells, insects, &c. in the said earths. There is a remarkable progression from the black alum slates to the pit-coal, in proportion as the quantity of the phlogiston encreases, and the quantity of the earth decreases (Sect. clix.). It is this phlogiston which makes this alum slate capable of burning by itself, when it is once lighted ; wherein it differs from the alum ores of Lumini, which, in order to be brought to moulder, require the being exposed to the heat of the sun, and to be sprinkled with water : the former has also within itself sufficient matter to spontaneously flame upon certain occasions, according to what the

celebrated

a. With the alcali of the common falt, or fea-falt, *Alcali minerali faturatum : Sal mirabile Glauberi.*

This is a neutral falt, prepared by nature, as well as by art, containing more or lefs of iron, or of a calcareous earth, from which arifes alfo fome difference in its effects, when internally ufed. It fhoots eafily into prifmatical criftals, which become larger in proportion to the quantity of water evaporated before the criftallifation. When laid on a piece of burning charcoal, or elfe burnt with a phlogifton, the vitriolic acid difcovers itfelf by the fmell like to the *hepar fulphuris.*

It is found in a diffolved ftate in fprings and wells, and in a dry form on walls, in fuch places where aphronitrum has efflorefced through them, and the vitriolic acid has happened to be prefent; for inftance, where marcafites are roafted in the open air. This falt is often confounded with the aphronitrum, or a pure mineral alcali; and a learned difpute once arofe, which of thefe falts ought with the greateft propriety to be called natron, *Baurach veterum, fal mirabile,* or Epfom falt; whereas it might eafily have been decided by chemical ex-

celebrated experiments of Lemery, and others, demonftrate, and from which caufe many volcanos and earthquakes may perhaps be deduced. The *pulvis pyrophorus* is alfo made of alum, intimately united with a phlogiftic fubftance ; in the preparing of which, they ought carefully to avoid that any iron enters the mixture, becaufe the acid has too ftrong an attraction to the iron, and cannot unite with the phlogifton alone, which, however, is quite neceffary in this operation.

periments,

periments, if their qualities had been regarded, in preference to their figures, or their native places.

This may be called Englifh or Epfom falt, when it has naturally as equal a copious portion of the calcareous earth as of the artificial one; but I have, in regard to its effects, for which it has been moft valued by Glauber, ranked all the lefs confiderable varieties of this neutral falt, when natural, under the name of *fal mirabile*.

SECT. CXXVII.

B. Acid of common or fea-falt, *Acidum falis communis.*

This acid, confidered in that ftate in which it can be had, viz. in mixture with water, has the following qualities.

1. It does not alter the fluidity of water, nor confiderably augment its heavinefs, as the vitriolic acid does.

2. It is fomewhat lefs corrofive and four than the faid vitriolic acid.

3. It ftrongly attracts the alcaline falts; but, however, is forced to quit them to the vitriolic acid, when that is added.

4. It diffolves the calcareous earth, and makes with it a fubftance, called *fal ammoniacum fixum.*

5. When expofed to the fire, combined with a phlogifton, it burns with a yellowifh green flame.

6. When highly concentrated and pure, as when it is diftilled from common falt mixed with pipe clay, it diffolves tin and lead: but

but lefs pure, it diſſolves copper, iron, zink, and the regulus of antimony : the copper is however more eaſily diſſolved, when it is in form of a calx, as the calces of quickſilver and cobalt likewiſe are.

7. It unites with ſilver diſſolved in aqua-fortis, and with lead diſſolved in aqua-regia, falling with them to the bottom, in form of a white ſpongy maſs. This preci-pitation, expoſed to the fire, ſtill retains the acid, and melts with it into a glaſſy ſubſtance, which does not diſſolve in water.

8. It is apt to attract the humidity of the air, and to promote the decaying of thoſe dry ſubſtances, with which it has been united.

9. Mixed with the ſpirit of nitre, it makes the ſo called aqua-regia, which is the true liquid menſtruum for gold.

This acid ſeems alſo, on certain occa-ſions, to have got looſe from thoſe ſub-ſtances, with which it has been originally united in the earth : the *ſal ammoniacum naturale* at Solfatara in Italy, and the horn ſilver ore (clxxvii.) appear to be proofs of this, as they ſeem to be the products of time.

SECT. CXXVIII.

1. Mixed or ſatiated acid of ſea-ſalt, *Acidum ſalis heterogeneis ſaturatum.*
A. With earths, *Terris ſaturatum.*
   1. With a calcareous earth, *Terrâ calcareâ ſaturatum : Sal ammoniacum fixum.*
      This ſomewhat decays or attracts the humidity of the air : it is found in abundance

abundance in the fea-water. See the calcareous kind, Sect. xxi.

SECT. CXXIX.

B. With alcaline falts, *Salibus alcalinis fatu-ratum.*

1. With the fixed mineral alcali, or fea alcali, common falt, or fea-falt, *Sal commune.*

This fhoots into cubical criftals during the very evaporation, it crackles in the fire, and attracts the humidity of the air.

a. Rock falt, foffil falt, *Sal montanum.*

Occurs in form of folid ftrata in the earth.

1. With fcaly and irregular particles, *Sal montanum particulis indeterminatis.*

  a. Grey, and

  b. White. Thefe are the moft common, but the following are fcarcer.

  c. Red,

  d. Blue, and

  e. Yellow, from Cracow in Poland, England, Salzberg, and Tirol.

2. Criftallifed rock falt, *Sal montanum cryftallifatum. Sal gemmæ.*

  a. Tranfparent, from Cracow in Poland, &c.

SECT. CXXX.

b. Sea Salt, *Sal marinum.*

Is produced from fea-water, or from the water of falt lakes, by evaporation in the fun, or by boiling.

The

The seas contain this salt, though more or less in different parts. In Siberia and Tartary there are lakes that contain great quantities of salt.

SECT. CXXXI.

c. Spring salt, *Sal fontanum.*

Is produced by boiling the water of the fountains near Halle in Germany, and other places. Near the city of Lidkoping, in the province of Westergottland, and in the province of Dal, salt-springs are found, but they contain very little salt: and such weak water is called *solen* by the Swedes *.

SECT. CXXXII.

2. Saturated with a volatile alcali, *Acidum salis communis alcali volatili saturatum.* Native sal ammoniac, *Sal ammoniacum naturale.*

This is of a yellowish colour, and is sublimed from the flaming fents or crevices at the Solfatara near Naples. See Sect. cxli.

---

* This division of the natural common salts is generally adopted, and not without reason, since the taste of all differs a little from one another, which depends on the less or greater mixture with heterogeneous substances. For out of the purest of these salts, a little of an earthy substance may still be precipitated, which dissolves in acids, and seems to be of a calcareous nature. The naturalists have troubled themselves a great deal to find out, how common salt is produced in the earth, and from whence the great store of it in the ocean is supplied: but they have proposed nothing but conjectures, without any wise illustrating the main question.

SECT.

SECT. CXXXIII.

C. United with phlogifton, *Acidum alis communis phlogifto faturatum.* Amber, *Succinum.* See Sect. cxlvi. *

SECT. CXXXIV.

D. United with metals, *Acidum falis metallis faturatum.*
1. With filver, *Acidum falis communis argento faturatum. Minera argenti cornea,*
Horn filver ore. The *Hornertz* of the
Germans. See Sect. clxxvii.

SECT. CXXXV.

The Second Order.

Alcaline Mineral Salts, *Alcalia Mineralia.*
Thefe are known by their action on the
above-mentioned acids, when they are joined
together, whereby a fermentation arifes, and
a precipitation enfues of fuch bodies as either
of them had before kept in diffolution;

---

* The dry volatile falt of amber, which difcovers itfelf to
poffefs the qualities of an acid, is, according to Mr. Bourdelin's experiments, communicated to the French Academy,
compounded of the acid of common falt, and a phlogifton,
both which fubftances are faid likewife to make out the conftituent parts of the yellow amber itfelf, though in different
proportion than in the falt: for this reafon, and until this
opinion is refuted by other experiments, the falt of amber
cannot be confidered as a mineral falt, that is different from
the others, and confequently exiftent by itfelf; nor can the vitriolic acid be faid to coagulate the yellow amber.

uniting

uniting at the fame time together, by which
new compofitions are made, that are called
neutral falts, or *falia neutra.*

Thefe alcaline falts are;

S E C T. CXXXVI.

1. Fixed in the fire, *Alcalia mineralia fixa.*
A. Alcali of the fea, or common falt, *Alcali
falis communis, propriè minerale dictum.*
 1. Pure, *Purum.*

This has nearly the fame qualities
with the lixivious falt, which is prepared from the afhes of burnt vegetables ; it is the fame with the *fal fodæ,*
or kelp, becaufe the kelp is nothing
elfe than the afhes remaining after the
burning of certain herbs that abound in
common falt; but which common falt,
during the burning of thofe vegetables,
has quitted its acid.

This,
1. Ferments with acids, and unites with
 them.
2. Turns the fyrup of violets to a green
 colour.
3. Precipitates fublimate mercury in an
 orange-coloured powder.
4. Unites with fat fubftances to make
 foap.
5. Diffolves the filiceous earth in the
 fire, and makes glafs with it, &c.
 It diftinguifhes itfelf from the falt of
 the pot-afhes, by the following properties : that,
6. It fhoots eafily into prifmatical criftals, that

7. Fall

7. Fall to powder in the air, which is effected by nothing elfe, than that they eafily lofe their humidity.

8. Mixed with the vitriolic acid, it makes the *fal mirabile*.

9. It melts eafier, and is fitter for producing the *fal commune regeneratum*, *nitrum cubicum*, &c. Perhaps it is alfo more conveniently applied in the preparation of feveral medicines.

10. It is fomewhat volatile in the fire*.

SECT. CXXXVII.

2. Mixed with a fmall quantity of the calcareous earth, *Alcali falis communis terræ calcareæ parvâ portione combinatum. Aphronitrum*.

* This falt is not met with pure in Europe, but it is faid to be found in both the Indies, not only in great quantity, but likewife of a tolerable purity: it is there collected in form of an efflorefcence in the extenfive deferts, a profitable trade being carried on in it for the making of foap and glafs: and therefore it is very probable, that the antients meant this falt by their *natron*, or Baurach. The calcareous earth is fufpected either to contain this falt in its own compofition, or elfe to be able to generate it from itfelf: but this hypothefis cannot be demonftrated. It is more probable, that the heat of the fun under the equator, and in the countries on both fides of it, evaporates the humidity, and afterwards expels the acid out of fome common falt, which either is naturally mixed with the earth, or elfe has been depofited there through the means of certain decayed vegetables, that always attract this falt; becaufe an earth from Paleftine, which Dr. Haffelquift fome years ago fent to Sweden, as a matrix of the natron, did upon trial yield nothing but a common falt: and it might, perhaps, have been taken at fuch a great depth, that it had not yet fuffered any decompofition. But this matter wants to be more illuftrated by obfervations, which might be beft made in the Eaft-Indies, where the greateft quantity is to be had, and alfo by fome farther analyfes of the fubftance.

This

This is fo ftrongly united with the calcareous earth, that the latter enters with it into the very criftals of the falt : though by repeated folutions the earth is by degrees feparated from it, and falls to the bottom after every folution. It grows in form of white froft on walls, and under vaults, and in places where it cannot be wafhed away by the rain. When it contains any confiderable quantity of the calcareous earth, its criftals become rhomboidal, a figure which the calcareous earth often affumes in fhooting into criftals : but when it is purer, the criftals fhoot into a prifmatical figure. This is a circumftance which neceffarily muft confufe thofe who know the falts only by their figure, and fhews, at the fame time, how little certainty fuch external marks afford in a true diftinction of things. This falt is therefore very often confounded with the *fal mirabile.*

SECT. CXXXVIII.

3. Saturated with mineral acids, *Alcali falis communis acidis mineralibus ad faturitatem mixtum.* Neutral falts, *Salia media, falia neutra.*

　a. With the acid of fea-falt, common falt, fea-falt, *Sal commune.* Sect. cxxix.

　b. With the vitriolic acid, *Sal mirabile.* Sect. cxxvi.

SECT. CXXXIX.

Ḃ. Borax.

This is a peculiar alcaline falt, which is fuppofed to belong to the mineral kingdom, and cannot be otherwife defcribed, than that it is either fome unknown alcali, united with an earth; which is diffoluble in water; and vitrefcible; or an alcaline falt, which is fixed in the fire, and melts to a glafs, which glafs is afterwards diffoluble in water.

Many experiments have been made with it, in order to difcover its origin and conftituent parts, and therefore it is amply treated of in chemical books; and its following qualities are to be obferved.

1. It fwells and froths in the fire, as long as any humidity remains in it, but melts afterwards very eafily to a tranfparent glafs, which, as it has no attraction to the phlogifton, keeps itfelf in the form of a pearl on the charcoal, when melted with the blowpipe.

2. It changes the fyrup of violets into green; and precipitates the folution of allum, and that of metals; made with acids.

3. It unites with mineral acids to a neutral falt, which fhoots into very fine and fubtile hair-like criftals, and is called *fal fedativum*. In a certain compofition it is volatile; and mixed with *litmus*, or *fuccus heliotropii*, and the fyrup of violets, it difcovers marks both of an alcali and an acid.

4. When it has been united with the vitriolic acid and a phlogifton, no *hepar fulphuris* is produced.

L   5. After

5. After being refined, it shoots into irregular
figures : but the cryftals, which form themfelves after the firft operation, and are called *Tincal*, confift of flat octagonal prifms,
flat at the extremities, and with their angles cut off or truncated *.

* It is yet unknown of what fubftance the Eaft-Indians and
Chinefe prepare the borax. The unrefined, which is brought
to Europe under the name of *tincal*, looks like foft foap, is
fat, and covers or encrufts the borax criftals. The mine-
mafter Mr. Swab, who has had an opportunity of making
experiments upon this *tincal*, has publifhed them in the Acts
of the Royal Academy of Sciences at Stockholm for 1756. He
fays, that he has found in it a martial earth, and a fat fubftance, which, to fmell and other circumftances, comes neareft
to a mineral fat : as likewife, that pure borax does not yield
any *hepar fulphuris*, when united with a phlogifton and a vitriolic acid; from which he concludes, that borax is prepared
from its own particular mineral fubftance.

Profeffor Pott and Mr. d'Henouville have very carefully
examined the refined borax ; and from their experiments,
which have been publifhed, it is evident, that it is of a particular alcaline nature : however, there yet remains to know
for certain, from what it is prepared by the Indians : for, if
it is produced from a mineral fubftance, as is very probable,
there muft exift other mixtures and compofitions, which are
yet unknown to the learned world.

I have alfo found in the *tincal* fmall bits of leather, bones
and fmall pebbles, whence there is no certainty to be concluded on from its examination ; but, if it fhould happen,
that it is prepared from animal fubftances, it muft be allowed,
that nature has formed an alcaline falt in the animal kingdom, which anfwers to the fixed acid falt in the human urine,
called *fal fufibile microcofmicum*, and which has been firft accurately defcribed by Mr. Margraff, in the Memoirs of the
Academy of Berlin.

Some years ago a report was propagated from Saxony, that
fomebody had there difcovered a fubftance out of which borax
could be made, and alfo the art of preparing it : but nothing
more has ever tranfpired fince, than that the author fhewed it
in fecret to his friends, and gave a defcription of it, which
only was intended to miflead them, if he really did poffefs
the art.

SECT.

SECT. CXL.

2. Volatile alkali, *Alcali minerale volatile*.

This perfectly refembles that falt which is extracted from animals and vegetables, under the name of *alcali volatile*, or *fal urinofum*, and is commonly confidered as not belonging to the mineral kingdom; but fince it is difcovered not only in moft part of the clays, but likewife in the fublimations at Solfatara near Naples, it cannot poffibly be quite excluded from the mineral kingdom.

Its principal qualities are that,

a. In the fire it rifes in *forma ficca*, and volatilifes in the air, in form of corrofive vapours, which are offenfive to the eyes and nofe.

b. It precipitates the folution of the mercurial fublimate into a white powder.

c. It alfo precipitates gold out of *aqua regia*, and detonates with it, becaufe

d. It has a reaction in regard to the acids, though not fo ftrongly as other alcalies.

e. It tinges the folution of copper blue, and diffolves this metal afrefh, if a great quantity is added.

f. It deflagrates with nitre, which proves that it contains a phlogifton.

It is never found pure, but

SECT. CXLI.

A. Mixed with

1. Salts, *Alcali minerale volatile falibus mixtum*.

L 2   *a.* With

a. With the acid of common falt, *Alcali minerale volatile acido falis unitum.* Native fal ammoniac, *Sal ammoniacum nativum,* Sect. cxxxii *.

SECT. CXLII.

2. With earths.

    a. Clay, *Alcali minerale volatile argillâ mixtum.*

The greateft part of the clays contain a volatile alcali, which difcovers itfelf in the diftillation of the fpirit of fea-falt, &c. †

---

* If that hypothefis could be proved true, which holds that volcanos and fubterranean fires arife from flates, formed from vegetables, animals, and the *humus ater* or mould, mixed together, (Sect. cxxiv.) the origin of the fal ammoniac at Solfatára would eafily be acknowledged ; fince we know that petrifactions difcover a principle within them, which contains the *fal urinofum.*

† In cafe fome of the clays are produced from the mould or *humus ater* (Sect. xci.) it is not difficult to fee the reafon of the prefence of this alcali in them ; but, though it would be both ufeful and curious to know all the changes of minerals, yet it is much better to take and employ them in their prefent ftate, than to lead the mind into perplexities by examining the combinations of thefe things by other means than by what the external fenfes fhew, and by rational experiments.

A German author has lately afferted, that metals, according to his experiments, have been found diffolved or mineralifed by a volatile alcali ; but, befides that fome fubjects mentioned by him ; for inftance, the *Sinople,* or Red Chalk ; the *Hungarian Gilben,* or Vein Stones ; and the *Horn Silver ore* ; do not fhew the leaft mark of it ; there is alfo wanted a defcription of the experiments he has made, and of the phœnomena which have prefented themfelves to him, during the examination of the other ores which he has mentioned : for thefe reafons his opinion cannot yet be admitted.

SECT.

SECT. CXLIII.

Observations on Salts.

The perfect knowledge of thefe bodies muft be had from chemical books and practical chemiftry, being almoft the chief fubject of that fcience. From thence we likewife learn why the acids are confidered as falts, though a certain figure neither is found, nor can reafonably be expected in them. We are farther taught, that the angular figure, which is fuppofed to be effential to falts, and by its varieties to mark out their different fpecies, depends on an alcali, earth, and metals, united with more or lefs water : for elfe, if this was not fo, the criftal of alum and vitriol ought to be of the fame figure, no *nitrum cubicum* would exift, nor could any criftallifation happen in fuch cafes, where the acids neceffarily muft be parted (Sect. xi.).

Salts are contained in all the three kingdoms of nature; and as it is not yet known how the changes happen, and how far the varieties depend on one another, we cannot attribute to the mineral kingdom any other falts than thofe which are found truly changed in the earth.

The ufe of the falts in medicine and in common life is fo great, that it would require a feparate treatife, if it were to be fully difcuffed. Mean while, every one who applies himfelf to the ftudy of mineralogy, in order to learn the ufe to which the mineral bodies can be employed in common life, I mean in particular mines, muft endeavour to difcover where falts may be found, and how they muft be prepared, fo as to be beft fit for ufe. But the preparation of falts is not the fubject of

L 3   this

this work; they are here only defcribed, fuch as they are naturally found, *viz.* entangled in certain heterogeneous bodies, of which they require but very little for their faturation.

SECT. CXLIV.

The THIRD CLASS.

MINERAL INFLAMMABLE SUBSTANCES,

Phlogifta Mineralia.

To this clafs belong all thofe fubterraneous bodies that are diffoluble in oils, but not in water, which they repel; catch flame in the fire; and are electrical.

It is difficult to determine what conftitutes the difference between the purer forts of this clafs, fince they all muft be tried by fire, in which they all yield the fame product; but thofe which in the fire fhew their differences by containing different fubftances, are here confidered as being mixed with heterogeneous bodies: that fmall quantity of earthy fubftance, which all phlogifta leave behind in the fire, is, however, not attended to.

SECT. CXLV.

1. Amber grife, *Ambra grifca.*

Is commonly reckoned to belong to the mineral kingdom, although it is faid to have doubtful marks of its origin.

   a. It has an agreeable fmell, chiefly when burnt.

   b. It is confumed in an open fire.

                                                    c. It

c. It foftens in a common degree of heat, fo as to ftick to the teeth like pitch.

d. It is of a black or grey colour, and of a dull and fine-grained texture. The grey is reckoned the beft, and is fold very dear.

This drug is brought to Europe from the Indies; it is employed in medicine, and as a perfume.

SECT. CXLVI.

2. Amber, *Ambra flava*; *Succinum*, *Electrum*.

This is a fubftance which is dug out of the earth, and found on the fea-coafts. According to the experiments of Mr. Bourdelin, it confifts of an inflammable fubftance, united with the acid of common falt, which feems to have given it its hardnefs. It is fuppofed to be of vegetable origin, fince it is faid to be found together with wood in the earth. By diftillation it yields water, oil, and a volatile falt, which the abovementioned author has found to be the acid of common falt, united with a little of the phlogifton. There are often found fifh, infects, and vegetables included in it, which teftify its once having been liquid. It is more tranfparent than moft part of the other bitumens, and is doubtlefs that fubftance which firft gave rife to electrical experiments.

Its varieties are reckoned from the colour and tranfparency: it is found

A. Opake, *Succinum opacum*.

   a. Brown.

   b. White.

   c. Blackifh.

B. Tranf-

B. Tranſparent, *Succinum diaphanum.*
  a. Colourleſs.
  b. Yellow.

    The greateſt quantity of European amber is found in Pruſſia; but it is, beſides, collected on the ſea-coaſt of the province of Skone, and at Biorko, in the Lake Malaren, in the province of Upland; as alſo in France and in Siberia. It is chiefly employed in medicines, and for making varniſhes.

SECT. CXLVII.

3. Rock-oil, *Petroleum.*

    It is an inflammable mineral, of a light-brown colour, which cannot be decompounded, but is often rendered impure by heterogeneous admixtures. In length of time, it hardens in the open air, like a vegetable reſin, and then becomes of a black colour, whether it is pure, or mixed with other bodies. It is likewiſe found in the earth

A. Liquid.
  1. Naphta, *Naphta.*

    This is ſaid to be of a very fragrant ſmell, tranſparent, extremely inflammable, and attracts gold. It is gathered from the ſurface of the water in ſome wells in Perſia.

SECT. CXLVIII.

2. Rock-oil, *Petroleum, propriè ſic dictum.*

    This ſmells like the oil of amber, though more agreeable, and is likewiſe very ready to take fire. It is collected

in

in the fame manner as the Naphta, from fome wells in Italy, and in a deferted mine at Ofmundfberget in the province of Dalarne: at this laft-mentioned place it is found in fmall hollows in the lime-ftone, as refin is in the wood of the pines.

SECT. CXLIX.

B. Thick and pitchy Rock-oil, or Barbadoes Tar, *Petroleum tenax, Maltha.*
This refembles foft pitch.
It is found in Moffgrufvan, at Norberg, in the province of Weftmanland, and at the Dead Sea in the Holy Land.

SECT. CL.

C. Hardened Rock-oil, *Petroleum induratum.* Foffil Pitch, *Pix montana.*
1. Pure, *Afphaltum.*
This leaves no afh or earthy fubftance when it is burnt.
It is found at Finnberget, in the parifh of Grythytta, in Weftmanland.
From this or the preceding fubftance, it is probable, the afphaltum was prepared that the Egyptians ufed in embalming their dead bodies, and which is now called *Mummia.*
2. Impure, *Pix montana impura.*
This contains a great quantity of earthy matter, which is left in the retort after diftillation, or upon the piece of charcoal, if burnt in an open fire; it coheres like a flag, and is of the colour of black lead: but in a calcining heat
this

this earth quickly volatilifes; fo that the nature of it is not yet known.

It is found in Moffgrufvan at Norberg, and in Grengierberget, both in the province of Weftmanland, and alfo in other places *.

SECT. CLI.

4. Mineral Phlogifton, or Bitumen, united with the vitriolic acid, *Phlogifton minerale acido vitrioli junctum.* Sulphur or Brimftone, *Sulphur.*

This is very common in the earth, and difcovers itfelf in many and various forms. It is found,

A. Native Sulphur, *Sulphur nativum.*

In this the two conftituent parts are mixed in due proportion in regard to each other, according to the rules of that attraction which is between them; it is eafily known,

1. By its inflammability, and by its flame.
2. By its fmell, when burnt; and,
3. By its producing a liver of fulphur, when mixed with a fixed alkali, like that made from artificial fulphur.

It is found

a. Pellucid, of a deep yellow colour.

b. Opake, white and greyifh.

Thefe are found in Siberia, at Bevieux in Switzerland, and at Solfatara near Naples. It is often found on limeftone, which the vitriolic acid has left untouched, having a ftronger attraction to

---

* The fubftance which rifes, and then falls into the receiver during the diftillation of this foffil pitch, is entirely the fame as the common natural liquid rock-oil, Sect. cxlviii.

the

the phlogifton, and therefore wholly
uniting with that.

SECT. CLII.

B. Sulphur that has diffolved, or is faturated
with metals, *Sulphur metallis faturatum.*
  1. With iron, *Sulphur marte faturatum.*
  Pyrites, or Copperas-ftone, *Pyrites.*
    This is the fubftance from which moft
  fulphur is prepared, and is therefore
  ranked here with all its varieties. It is
  hard, and of a metallic fhining colour.
  a. Pale yellow Pyrites, *Pyrites fubflavus.*
  Marcafite.
    This is very common, and contains
  a proportionable quantity of fulphur
  with refpect to the iron; when once
  thoroughly inflamed, it burns by itfelf.
    1. Of a compact texture, *Texturâ
      æquali : PolitaPiedra del Ynca, Hif-
      panorum.*
    2. Steel-grained, *Texturâ chalybeâ.*
    3. Coarfe-grained, *Texturâ granulatâ.*
    4. Criftallifed, *Chriftallifatus.*
      It fhoots moftly into cubical and
    octoedral figures, though it alfo
    criftallifes into innumerable other
    forms.

SECT. CLIII.

b. Liver-coloured Marcafite, *Pyrites colore rubefcente.*
  Its colour cannot be defcribed, being betwixt that of the preceding
marçafite, and the azure copper ore.
                                When

When it is of a light colour, it is called in Swedish *Tennbett*, or *Wattnkies*, but *Lefverflag* when it is of a deeper colour.

The iron prevails in this kind; it is therefore leſs fit to have ſulphur extracted from it, and alſo for the ſmelting of copper ores. It is found

1. Of a compact texture, from Nya Kopparberget in the province of Weſtmanland.

2. Steel-grained, from Stollberget in Weſtmanland.

3. Coarſe-grained, from Weſter-ſilfverberget in Weſtmanland.

SECT. CLIV.

2. Iron and Tin, *Sulphur ferro & ſtanno ſaturatum*. Black Lead, or Wadd, *Molybdæna*.

If by ſuch a mixture as this the iron and tin be not rendered too volatile, it muſt be ſuppoſed that the great loſs the Black Lead ſuſtains in the calcining heat is occaſioned from the ſulphur, and that the ſulphur conſequently makes out the greateſt part of the black lead. It is found,

a. Lamellar and ſhining, of the ſame colour as the potters lead ore, *Molybdæna membranacea nitens.*

From Biſpergs Klack in the province of Dalarne, Baſtnas-grufva at Riddarſhyttan in Weſtmanland, Altenburg in Saxony.

The

'The variety from Bifpergs Klack has been examined by Mr. Quift, and has, by its volatilifing under the muffel, in form of a white fibrous fublimate, induced that gentleman to examine the black lead more particularly; and he has publifhed fome very remarkable experiments on it in the Tranfactions of the Academy of Sciences at Stockholm, for the year 1754.

b. Of a fteel-grained and dull texture, *Texturâ chalybeâ.* It is naturally black, but when rubbed it gives a dark lead colour.

c. Of a fine fcaly and coarfe-grained texture, *Texturâ micaceâ & granulatâ.* Coarfe Black Lead.

It has at the fame time a fcaly and a granulated appearance.

From Gran in the province of Upland, and from Tavaftehuflan in Finland *.

* Profeffor Pott has examined the black lead in covered veffels, and Mr. Quift in an open fire, from which difference in the method of treating it, different notions have arofe: becaufe the black lead is nearly unalterable when expofed to the fire in covered veffels, or when immediately put into a ftrong charcoal fire, but it is almoft wholly volatile in a calcining heat. This is the cafe with feveral others of the mineral phlogiftons; and from this we may in general learn, how neceffary it is to examine the mineral bodies by many and different methods, and to endeavour to multiply the experiments more than what has been hitherto done. Pencils are made from the black lead; as alfo the black lead crucibles.

SECT.

SECT. CLV.

3. Sulphur with iron and copper, yellow or marcafitical copper ore. See Sect. cxcviii.
4. Sulphur with iron and lead, Potters lead ore. See Sect. clxxxix.
5. Sulphur with iron and zinc, mock lead, black jack, or blende. See Sect. ccxxix.
6. Sulphur with iron and arfenic, arfenical pyrites. Sect. ccxliii.
7. Sulphur with iron and cobalt. Sect. ccl.
8. Sulphur with iron and bifmuth. Sect. ccxxv.
9. Sulphur with iron and nickel. Sect. cclvi.
10. Sulphur with iron and gold, pyritical gold ore. Sect. clxvi.

SECT. CLVI.

11. Sulphur with filver, glafs filver ore. Sect. clxix.
12. Sulphur with copper, grey or vitreous copper ore. Sect. cxcvii.
13. Sulphur with lead, Potters lead ore. Sect. clxxxvii.
14. Sulphur with bifmuth. Sect. ccxxiv.
15. Sulphur with quickfilver, cinnabar. Sect. ccxviii.
16. Sulphur with arfenic, Orpiment, Realgar. Sect. ccxli.

SECT.

SECT. CLVII.

5. Mineral phlogifton united with earths, *Phlogifton minerale terris imbutum.*
A. With a calcareous earth, *Phlogifton minerale terrâ calcareâ imbutum.*
   1. With pure calcareous earth, the fœtid or fwine fpar, Sect. xxiii.
   2. With the calcareous earth and vitriolic acid, the Leberftein or Liverftone of the Swedes, Sect. xxiv.

SECT. CLVIII.

B. With an argillaceous earth, *Phlogifton argillâ mixtum.*
   1. With a fmall quantity of argillaceous earth and vitriolic acid: Coal: Lithantrax.

   It is of a black colour, and of a fhining texture; it burns, and is moftly confumed in the fire; but leaves however a fmall quantity of afhes.
   a. Solid coal.
   b. Slatty coal
   Found in England, and at Boferup in the province of Skone.

SECT. CLIX.

   2. With a greater quantity of argillaceous earth and vitriolic acid, the Kolm of the Swedes. This is of the fame appearance with the former, though of a more dull texture; it burns with a flame, and yet is not confumed, but leaves behind a flag
of

of the fame bulk or volume as the coal was.

From England, and among the alum rock at Moltorp and Billingen in the province of Weftergottland.

S E C T. CLX.

3. With abundance of argillaceous earth, Stone coal. It burns with a flame by itfelf, otherwife it looks like other flates.

It is found at Gullerafen, in the parifh of Rettwik, in the province of Dalarne, and alfo with the coals at Boferup in Skone *,

S E C T. CLXI.

6. Mineral phlogifton mixed with metallic earths, *Phlogifton minerale metallis impregnatum.*

This is not found in any great quantity: in regard to its external appearance, it refembles pit-coal; and the fat fubftance contained in it, at times partly burns to coal, and partly volatilizes in a calcining heat.

The only known varieties of this kind are, *A. Minera cupri phlogiftica.*

When it has been inflamed, it retains the fire, and at laft burns to afhes, out of which

---

* This laft mentioned kind has induced me to believe, that the earth of the pit-coals is an argillaceous one, but is not fo eafy to be difcovered after its being burnt. The pit-coals contain more or lefs of the vitriolic acid, for which reafon the fmoak arifing from them attacks filver in the fame manner as fulphur does; though the coals be ever fo free from marcafite, which however is often found imbedded or mixed with them.

pure

pure copper can be fmelted. It is found in Sladkierr's Grufva in the province of Dal, and at Bifpergs Klack, in the province of Dalarne.

B. *Minera ferri phlogiftica.*

This is not very different in its appearance from the pit-coal or foffil pitch; but it is fomewhat harder to the touch: there are two varieties of this fpecies:

1. Fixt in the fire, *Minera ferri phlogiftica fixa.*

Expofed to a calcining heat, it burns with a very languid though quick flame, it preferves its bulk, and lofes only a little of its weight. It yields above 30 per cent. of iron.

a. Solid, refembles black fealing-wax.

It is found in the liver-coloured marcafite, (Sect. cliii.) in Wafkberget, at Norrberke, in Weftmanland.

b. Cracked, and friable, from Finnberget, at Grythyttan, in Weftmanland.

2. Volatile in the fire, *Minera ferri phlogiftica volatilis.*

This is unalterable in an open fire, either of charcoal, or even upon a piece of charcoal before the flame of the blowpipe: but under a muffel the greateft part of it volatilifes, fo that only a fmall quantity of calx of iron remains. It is found,

a. Solid, from Kronprints Shurff, at Kongfberg, in Norway.

b. Cracked, from the parifh of Quiftbro, in the province of Nerike.

This laft kind leaves more afhes: thefe afhes, when farther expofed to

M the

the fire, become firſt yellowiſh green,
and afterwards reddiſh brown, when,
beſides iron, they then alſo diſcover
ſome marks of copper; it has however not been poſſible to extract any
metallic ſubſtance from them, the
effects of the loadſtone, and the colour communicated to the glaſs of
borax, having only given occaſion to
this ſuſpicion (conf. Sect. cl. & cliv.).

S E C T.   CLXII.

Observation on the Bitumens.

That ſubſtance which the chemiſts call Phlogiſton, or an inflammable principle, exiſts in moſt
of the mineral bodies, though often in ſo ſmall a
quantity as not to be perceived; and therefore I
have here only enumerated thoſe kinds in which
it exiſts as a principal character; for inſtance, in
the fœtid ſpar or ſwine-ſtone, &c.

I do not myſelf know the ſubſtance in its ſimple
ſtate which I call a mineral phlogiſton, ſince the
ambergriſe and the rock-oil can be nothing elſe
than compoſitions which cannot be perfectly decompounded; and beſides, they are not to be extracted from coal, ſulphur, &c. which yet contain an inflammable ſubſtance. It ſeems as if a
great part of this claſs were originally generated
from the animal and vegetable kingdoms; ſo that
they firſt have been an *humus ater* or mould, with
which a vitriolic acid has afterwards been mixed;
and that they have been beſt able to retain this
phlogiſton, when they have been covered and
joined together by another earth: the coal, coal ore,
and pitch turff, (Sect. ccxciii.) give ſome hints or
reaſons

reafons for this fuppofition. The generation of fulphur and marcafite requires no preferable phlogifton out of any of the kingdoms of Nature, for the phlogiftons throughout all nature are equally alike fit for it.

It is a fublime fubject for philofophers to enquire how far fire, phlogifton, and electricity, have an affinity with, or dependance on, one another; but as they yet want that light in this matter which they wifh to have, I hope to be excufed for not mentioning any theories on the fubject.

This clafs is of great ufe in medicine; for inftance, the ambergrife, the falt of the yellow amber, the rock-oil, the afphaltum, and the fulphur. The rock-oil and fulphur are ufed in fireworks, the afphaltum by the watchmakers, and the yellow amber is ufed by the varnifhers and painters *.

SECT. CLXIII.

The FOURTH CLASS.

METALS, *METALLA,*

Are thofe mineral bodies which, with refpect to their volume, are the heavieft of all hitherto-known bodies; they are not only malleable, but they may alfo be decompounded, and in a melting heat be brought again to their former ftate,

---

* The coals, however, are of the greateft confequence for their œconomical ufe; and happy therefore are thofe countries which have a fufficient quantity of them, fince they may be employed as fuel to almoft every purpofe, which is plainly proved in England. E.

by

by the addition of the phlogiston they had lost in
their decompofition *.

S E C T. CLXIV.

The First Order.

Metals, *Metalla*.

1. Gold, *Aurum, Sol Chymicorum.*
This is by mankind efteemed as the principal and firft among the metals; and that
partly for its fcarcity, but chiefly for its following qualities.
1. It is of a yellow fhining colour.
2. It is the heavieft of all known bodies, its
  fpecific gravity to water being as 19,640
  to 1000.
3. It is the moft tough and ductile of all
  metals; becaufe one grain of it may be
  ftretched out fo as to cover a filver wire of

* Thofe metals which in a calcining heat lofe their phlogiston, and confequently with that the former coherency of
their particles, are called *imperfect*, as tin, lead, copper, and
iron, and all the femi-metals (of which more hereafter) : not-
withstanding which they may be malleable. But thofe which
cannot be deftroyed in the fire alone are called *perfect*, as gold,
filver, and platina del pinto. Neverthelefs, the metals have
commonly been confidered more with regard to their mallea-
bility than to their fixity in the fire, and are therefore divided into,
  A. Malleable, which are called *metals* ; and
  B. Brittle; which are called *femi metals*.
  The zinc is, however, as a medium between thefe two
divifions, juft as the quickfilver is between the perfect
and imperfect metals, becaufe the quickfilver may indeed
be fo far deftroyed in the fire, that its particles are feparated during their volatilifation ; but every one of them,
even the minuteft, retains however the phlogiston united
with it.

the

the length of ninety-eight yards, by which
means $\frac{1}{705600}$ grain becomes vifible to the
naked eye.

4. Its foftnefs comes neareft to that of lead,
and confequently it is but very little elaftic.

5. It is fixed and unalterable in air, water,
and fire, becaufe it does not eafily quit its
phlogifton; its liquid menftruum (7") being only made by art.

It has, however, according to Homberg's experiments, when expofed to
Tfchirnhaufen's burning-glafs, been found
partly to volatilife in form of fmoke,
and partly to fcorify: But this wants to
be farther examined. It is alfo faid,
that gold in certain circumftances, and
by means of certain artifices in electrical experiments, may be forced into glafs;
and that on this occafion it becomes white,
leaving a black duft behind it; which, if
fo, confirms certain other chemical experiments; viz. That gold can, together
with its colour, lofe fomething of its phlogifton, and yet retain its heavinefs, ductility, &c.

6. When melted, it reflects a blueifh green
colour from its furface.

7. It diffolves in aqua regia, which is compofed of the acids of fea-falt and nitre; but
not in either alone, nor in any other folution of falt or acid whatfoever.

8. When mixed with a volatile alcali and a
little of the acid of nitre, by means of
precipitation out of aqua regia, it burns
off quickly, in the leaft degree of heat,
with a ftrong fulmination.

M 3

9. It

9. It is diſſolved, *in formâ ſiccâ*, by the liver of ſulphur, and alſo ſomewhat by the glaſs of biſmuth.

10. It is not carried away by the antimony during the volatiliſation of that ſemi-metal, and is therefore conveniently ſeparated from other metals by the help of crude antimony, in which proceſs the other metals are partly made volatile, and fly off with the antimony, and partly unite with the ſulphur, to which the gold has no attraction, unleſs by means of ſome uniting body, or by a long digeſtion.

11. The phoſphorus is ſaid to have ingreſs into gold.

12. If mixed with a leſs portion of ſilver, platina, copper, iron, and zinc, it preſerves tolerably well its ductility; but,

13. When mixed with tin it becomes very brittle; and it attracts likewiſe the ſmoke of that metal, ſo as to be ſpoiled, if melted in an hearth where tin has been lately melted: And this is perhaps the reaſon why gold becomes brittle, and of a paler colour, when melted in a new black lead crucible (Sect. cliv.)

14. It requires a ſtrong heat before it melts, nearly as much, or a little more than copper.

15. It mixes or amalgamates readily with quickſilver.

16. It is not diſſolved by the glaſs of lead, and therefore remains on the cuppel.

In conſequence of theſe its principal qualities, it ſeems as if it could never be found in the earth but in a native or pure

ſtate;

ftate ; there are, however, feveral inftances that it has been found diffolved or mine-ralifed.

SECT. CLXV.

A. Native Gold, *Aurum nativum*,
Is in its metallic form commonly pure : And in this ftate moft part of this metal ufed in the world is found. With refpect to either the figure or the quantity in which it is found in one place, it is by miners divided into,

1. Thin fuperficial plated or leaved gold, which confifts of very thin plates or leaves, like paper.
2. Solid or maffive, is found in form of thick pieces.
3. Criftallifed, confifts of an angular or crif-talline figure.
4. Wafh Gold, or Gold Duft, is wafhed out of fands, wherein it lies in form of loofe grains and lumps *.

* The gold is in general more frequently imbedded and mixed with quartz, than with any other kind of ftone ; and the quartz in which the gold is found in the Hungarian gold mines is of a peculiar appearance. All other forts of ftones, however, are not to be excluded, fince gold is likewife found in fome of them ; for inftance, in limeftone (Sect. ix.) in Adolph Fredrik's Grufva at Adelfors, in the province of Smo-land ; in Hornblende (Sect. lxxxviii.), in Baftnas Grufva at Riddarfhyttan, in the province of Wellmanland ; not to men-tion feveral foreign gold mines.

The greateft quantity of gold is imported into Europe from Chili and Peru, in America ; and a little from China, and the coaft of Africa. The chief European gold mines are thofe of Hungary, and next to them thofe at Saltzburg. Befides thefe, there are fome others of lefs confequence ; among which the gold mines at Adelfors in Smoland deferve to be taken much notice of, not only on account of the veins already worked, but alfo in regard to the vaft tract of land, within which new

veins

SECT. CLXVI.

B, Mineralifed Gold, *Aurum mineralifatum.*

This is an ore in which the gold is fo far mineralifed, or fo entangled in other bodies, as not to be diffolved by the aqua regia.

1. Mineralifed with fulphur, *Aurum fulphure mineralifatum.*

　　a. Mineralifed by means of iron, *Aurum fulphure mineralifatum mediante ferro.* Marcafitical gold ore, *Pyrites aureus.*

　　　　It is found at Adelfors, in the province of Smoland, and contains an ounce of gold, or lefs, in an hundred pounds.

　　b. Mineralifed by means of quickfilver, *Aurum fulphure mineralifatum mediante mercurio.*

　　　　It is found in Hungary.

　　c. Mineralifed by means of zink and iron, *Aurum fulphure mineralifatum mediante zinco & ferro, aut argento.* The Schemnitz blende.

　　　　At Schemnitz in Hungary are found zink ores, which contain a great deal of filver, and this filver is very rich in gold (Sect. clxxv.) *.

veins are daily difcovered. The filver from the mines at Ofterfilverberger, in the province of Dalarne, contains about a fourth part of an ounce of gold in every pound of filver. Some native gold has likewife been found in Swappawari, above Tornea in Lapland, and in Baftnas, near Riddarfhyttan in Weftmanland.

* Since gold and fulphur have no immifcible power or attraction to one another, many have infifted that gold never could be found in marcafi e, or thofe ores which contain fulphur: But fince we know by experience, that gold can be
melted

SECT. CLXVII.

2. Silver, *Argentum, Luna*; which is,
a. Of a white shining colour.
b. Its specific gravity to water is 11,091 to 1000.
c. It is very tough or ductile, so that a grain of it may be stretched out to three yards in length, and two inches in breadth.

melted out of the above-mentioned ores, altho' they have been previously digested in aqua regia ; and that gold likewise mixes and diffolves into a regulus; there is the greatest reason to believe that a third substance, which here is a metal, must necessarily have by its admixture enabled the fulphur to unite with a certain quantity of gold. Scheffer has given upon this subject some very curious and useful observations, in his History of the Refining of Metals, inserted in the Transactions of the Academy of Sciences at Stockholm. It is very remarkable that the Mine-Master Henckel, in his excellent Treatise *de Appropriatione*, should be so obstinate in denying that marcasite could contain a diffolved gold.

It is, however, by no means hereby intended to confirm the credulous in their opinion, that the marcasites in general contain more gold than what true metallurgists have afferted; because fraud might then perhaps become too common. It is only meant to indicate, that, as no gold is to be expected from marcasites, where no native gold is found in the neighbourhood, in the fame manner no marcasites ought to be despised, which are found in tracks where gold ores are dug ; but at the fame time care must be taken not to be deluded by the mention of volatile gold, as it is a notion really contradictory and suspicious, and then there can be no fear of being misled.

I am not perfectly clear, if the gold is really diffolved and indurated, or, if I may fo exprefs myself, vitrified in the Shirls (Schirlkorrern), provided by this mineral body is meant a garnet substance (Sect. lxviii.) But I have seen a piece of what is called *Shirl*, whose texture was exactly like the Schemnitz blende ; and in this cafe it might perhaps hold the fame contents (Sect. clxxv.) For the other gold ores, I have not had an opportunity of feeing any from those places where gold is fearched for and really found.

d. It

d. It is unalterable in air, water, and fire.

e. It diſſolves in the acid of nitre, and alſo by boiling in the acid of vitriol.

f. If precipitated out of the acid of nitre with the common ſalt, or with its acid, it unites ſo ſtrongly with this laſt acid, that it does not part from it, even in the fire itſelf, but melts with it into a maſs like glaſs, which is called *luna cornea.*

g. It does not unite with the ſemi-metal nickel, during the fuſion.

h. It amalgamates eaſily with quickſilver.

i. It is in the dry way diſſolved by the liver of ſulphur.

k. It has a ſtrong attraction to ſulphur, ſo as readily to take a reddiſh yellow or black colour, when it is expoſed to ſulphureous vapours.

l. It has no attraction to arſenic; whence when the red arſenical ſilver ore, or *Roth-gulden Ertz* of the Germans, is put into the fire, the arſenic flies off, and leaves the ſulphur (which in this compound was the *medium uniens*) behind, united with the ſilver in form of the glaſs ſilver ore, or glaſs ertz.

m. It is not diſſolved by the glaſs of lead, and conſequently it remains on the cuppel.

n. It is exhaled or carried off by volatile metals and acids, as by the vapours of antimony, zink, and the acid of common ſalt.

o. It melts eaſier than copper.

SECT.

SECT. CLXVIII.

Silver is found,

A. Native or pure, *Argentum purum nativum.*
Native filver moft generally is nearly of fixteen carats ftandard.

1. Thin fuperficial plated or leaved filver.
2. It is alfo found in form,

    a. Of fnaggs, and coarfe fibres.
    b. Of fine fibres. Capillary filver.
    c. Arborefcent. From Potofi in America, and Kongfberg in Norway.
    d. Criftalline, or figured. This is very fcarce to be met with: it has diftinct figures, with fhining furfaces; it is, however, fometimes found at Kongfberg.

The filver from America is faid to be found for the moft part native; fo it is likewife at Kongfberg in Norway, but it is not commonly fo in the other European mines. In Sweden it is found native in a very fmall quantity, in the mines of Salberg in Weftmanland, of Lofafen in Dalarne, of Hevaffwik and Sladkierr in the province of Dal, of Sunnerfkog in the province of Smoland, and in the Ifland Utoen in the Lake Malaren. It was once found in pretty large lumps in a vein of clay in one of the iron mines at Normark, in the province of Wermeland. It was there mixed with nickel, which was partly decayed or withered; and under this circumftance it formed the compound ore called the *Stercus Anferinum,* or Goofe-dung Ore. At this place the

argilla-

argillaceous vein croffes the veins of the iron ore, and will perhaps be found to have more of thefe riches, even in feveral other places, if well fearched, as is done in other countries, oftentimes not on fuch evident marks or figns.

SECT. CLXIX.

B. Diffolved and mineralifed, *Argentum mineralifatum.*

1. With fulphur alone, *Argentum fulphure mineralifatum.* Glafs filver ore, *Minera argenti vitrea.*

This is ductile, and of the fame colour as lead; but, however, becomes blacker in the air. It has, therefore, very undefervedly got the name of glafs ore, for that name rather belongs to the *minera argenti cornea,* or horn filver ore, if indeed any filver ore can be confidered as glaffy.

It is found in the fame manner as native gold; viz.

1. In crufts, plates, or leaves.
2. Grown into,
    a. Snaggs, and,
    b. Criftalline figures.

It is generally either of a lamellar or a grained texture, and is found at Kongfberg and in the Saxon mines.

The glafs filver ore is the richeft of all filver ores; fince the fulphur, which is united with the filver in this ore, makes out but a very fmall quantity of its weight.

SECT.

SECT. CLXX.

2. With sulphur and arsenic, *Argentum sulphure & arsenico mineralisatum.* *Minera argenti rubra*, The red or ruby-like silver ore. The *Rothgulden* of the Germans.

The colour of this ore varies as the proportion of each of these ingredients varies in the mixture; viz. from dark grey to deep red: but when it is rubbed or pounded, it always gives a red colour. When put in the fire, it crackles and breaks; and when the crackling ceases, it melts easily, the arsenic at the same time exhaling in smoke.

a. Grey arsenical silver ore; which is either,

1. Plated, crusted, or leaved, and,
2. Solid.

b. The red arsenical silver ore.

1. Plated, crusted, or leaved.
2. Solid or scaly, and,
3. Cristallised.

In this last form it shews the most beautiful red colour, and is often semi-transparent. It contains about sixty per cent. in silver; and is found in the greatest quantity at Andreasberg in the Hartz.

SECT. CLXXI.

3. With sulphurated arsenic and copper, *Argentum arsenico & cupro sulphurato mineralisatum.* *Minera argenti albi,* The *Weissgulden* of the Germans.

This, in its solid form, is of a light grey colour, and of a dull and steel-grained texture.

ture. The more copper it contains, the darker is the colour. It often holds feven pounds of filver per cent. It is,

a. Friable, withered, or decayed, of a black or footy colour, and is therefore by the Germans called *Silber-Schwartz*, or *Ruffigtes-Ertz.*

b. Solid, of a light-grey colour, and is that fort properly fo called Weiffgulden.

It is found at St. Mary of the mines in Alfatia, the Saxon mines, and at St. Andreafberg in the Hartz.

SECT. CLXXII.

4. With fulphurated arfenic and iron, *Argentum ferro & arfenico fulphurato mineralifatum*, The *Weifertz*, or white filver ore, of the Germans.

This is an arfenical pyrites, which contains filver; it occurs in the Saxon mines, and fo exactly refembles the common arfenical pyrites as not to be diftinguifhed from it by fight alone, or without other means. The filver it contains may perhaps confift of very fubtile capillary filver mixed in it. However, I have not had an opportunity to examine this circumftance.

SECT. CLXXIII.

5. With fulphurated antimony, *Argentum antimonio fulphurato mineralifatum.*

a. Of a dark-grey and fomewhat brownifh colour. The *Lebererz*, from Braunfdorff in Saxony.

b. Of a blackifh blue colour.

ı. In

1. In form of capillary criftals, *Minera argenti antimonialis capillaris.* *Federertz,* or plumofe filver ore.

It is found in Saxony, and contains only two or four ounces of filver per cent.

SECT. CLXXIV.

6. With fulphurated copper and antimony, *Argentum cupro & antimonio fulphurato mineralifatum.* The Dal *Falertz.*

This refembles, both in colour and texture, the dark-coloured Weiffgulden, or Falertz. When rubbed, it gives a red powder.

a. Solid.

b. Criftallifed, is found in the parifh of Aminfkog, in the province of Dal; and at that place has been for feveral years melted by a method invented for the 'different mixture of the ores; which procefs muft be very troublefome to thofe who are not perfectly well verfed in metallurgy.

It contains thirteen ounces of filver, and twenty-four per cent. of copper.

SECT. CLXXV.

7. With fulphurated zink, *Argentum zinco fulphurato mineralifatum.* The *Pechblende* of the Germans.

This is a zink ore, mock lead, or blende, which contains filver, and is found among rich filver and gold ores; for inftance, in the Hungarian and Saxon mines.

a. Of

a. Of a metallic changeable colour.
  1. Solid, and with fine fcales.
  2. In form of balls. The *Kugel-ertz,*
    or ball ore.
      It is found at Schemnitz, and con-
    tains alfo gold. Its yield of filver is
    twenty-four ounces per cent. and
    thirty per cent. of zink.

b. Black mock lead, or blende, found in
  Saxony. This is alfo found,
  1. Solid, and with fine fcales;
  2. And in form of balls.

SECT. CLXXVI.

8. With fulphurated lead, Potter's ore, *Galena, Bleyglanz.* See Sect. clxxxviii.
9. With fulphurated lead and antimony,
called *Striperz.* See Sect. cxc.
10. With fulphurated iron, *Argentum ferro
fulphurato mineralifatum. Silberhaltiger kies,*
Marcafite holding filver.

    At Kongfberg in Norway, it is faid, a
liver-coloured marcafite is often found,
particularly at the mine called Fraulein
Chriftiana, &c. This marcafite contains
of filver from three to three ounces and an
half per cent.

SECT. CLXXVII.

11. With the acid of common falt, *Argentum
acido falis folutum & mineralifatum. Minera
argenti cornea. Hernertz,* or horn filver ore.
  This is the fcarceft filver ore; it is of
a white or pearl colour, changeable or
varying on the furface, femi-tranfparent,
               and

and fomewhat duſtile, both when crude, and when melted. It cannot be decompofed without fome admixture of fuch fubftances as attract the acid of the fea-falt. It is found in very thin worked or wrought leaves or crufts, at Johan Georgenſtadt, in Saxony.

S E C T. CLXXVIII.

OBSERVATIONS on the SILVER ORES.

Silver may, perhaps, be found mineralifed in the like manner with other metals than thefe here enumerated, fuch as with cobalt and bifmuth; but having no certain knowledge of fuch mineralifations, I omit them here. It would be worthy examining, if in thofe mine countries where gold and filver are found in quantity, other ores do not contain a little of thofe metals, more efpecially when the particles of filver and gold have not been able to extricate themfelves from the other minerals, and lie feparate from them in the fiffures, veins, and fhakes or wranks, that is hollow places, in the mines.

Thofe filver ores which are named from earth or ſtones, wherein the filver is found; as, for inftance, in the Goofe-dung filver ore, and the *Leberertz*; ought no more to be confidered in a natural fyftem than other diftinctions which are ufed at mineral works, and are only names given to the ores, according to the feveral changes they undergo to make them fit for the melting procefs.

In this our time a mineralifation of filver with alcali has been mentioned: it is faid to have been found at Annaberg in Auftria: But this difcovery, which is made by a mine-mafter, Mr. Von Jufti,

N   requires

requires an explanation, fince the author in his defcription does not obferve the neceffary diftinction between alcali and lime; and quotes the horn filver ore, and the *luna cornea*, as proofs of his opinion; by which, however, his opinion feems rather weakened than confirmed.

SECT. CLXXIX.

3. Platina del Pinto, Platina di Pinto, *Juan blanca*.

This metal is a recent difcovery of our times, and is defcribed with great accuracy by Scheffer, in the Acts of the Royal Academy of Sciences at Stockholm, for the year 1752; as alfo by Dr. Lewis, in the Philofophical Tranfactions for the year 1754, vol. xlviii. And though thefe two gentlemen agree in the principal circumftances relating to this metal, yet it is very plain by their defcriptions, that neither of them knew any thing of the other's experiments. By thefe defcriptions we are convinced of the refemblance this metal bears to gold; and therefore we muft allow it to be called *white gold*, though, both theoretically and practically, it may be diftinguifhed from gold by the following qualities.

1. It is of a white colour.
2. It is fo refractory in the fire, that there is no degree of heat yet found by which it can be brought into fufion by itfelf, the burning-glafs excepted, which has not yet been tried. But, when mixed with other metals, and femi-metals, it melts very eafily, and efpecially with arfenic, both in its

its metallic form, and in form of a calx or glafs.

3. It does not amalgamate with quickfilver, by itfelf, but only by means of the acid of common falt after a long trituration. This metal is therefore really feparated from gold by amalgamation, at thofe places where it is found; and without this quality, it would be very difficult to feparate it.

4. It is harder and lefs coherent than gold.

5. It is heavier than gold; and therefore the heavieft of all bodies hitherto difcovered: For though the fpecific gravity of platina, in the hydroftatical experiments made by Dr. Lewis, is found to be to water only as 17,000 to 1000; yet, when melted with other certain metals, its fpecific gravity has, by an exact calculation, been found to be confiderably augmented, even fo much as to 22,000.

6. Diffolved in aqua regia, and precipitated with tin, or with a folution of that metal, it yields no *purpura mineralis*.

Except thefe, this metal has the fame qualities as gold; but it cannot, on account of its refractorinefs in the fire, be worked off pure on the cuppel, nor likewife can it be worked with antimony; becaufe, before it is rendered perfectly pure, it cools, grows hard, and retains always fome part of the added metals. It is brought to us only in its native ftate, in fmall, irregular, rugged grains; and it is yet uncertain whether it is found naturally mineralifed. The Platina is brought to

Europe

Europe from the Rio de Pinto, in the
Spanish Weft-Indies.

SECT. CLXXX.

4. Tin, *Stannum*, *Jupiter*.

This is diftinguifhed from the other metals
by its following charaćters and qualities.

a. Of a white colour, which verges more to
the blue than that of filver.

b. It is the moft fufible of all metals; and,

c. The leaft dućtile; that is, it cannot be
extended or hammered out fo much as the
others.

d. In breaking or bending it makes a crackling noife.

e. It has a fmell particular to itfelf, and which
cannot be defcribed.

f. In the fire it is eafily calcined to white
afhes, which are twenty-five per cent. heavier than the metal itfelf. During this
operation, the phlogifton is feen to burn
off in form of fmall fparkles among the
afhes, or calx.

g. This calx is very refraćtory; but may,
however, with a very ftrong degree of
heat, be brought to a glafs of the colour
of hard refin. But this calx is eafily mixed
in glafs compofitions, and makes with them
the white enamel.

h. It unites with all metals and femi-metals;
but renders moft of them very brittle, except lead, bifmuth, and zink.

i. It amalgamates eafily with quickfilver.

k. It diffolves in aqua regia, the fpirit of fea-
falt, and the vitriolic acid; but it is only
corroded

corroded into a white powder by the spirit of nitre.

The vegetable acid, soaps, and pure alcaline salts, also corrode this metal by degrees.

l. Its specific gravity to water is as 7400 to 1000, or as 7321 to 1000.

m. Dissolved in aqua regia, which for this purpose ought to consist of equal parts of the spirit of nitre and sea-salt, it heightens the colour of the cochineal, and makes it deeper; for otherwise that dye would be violet.

SECT. CLXXXI.

Tin is not found naturally in the earth in any other state than,

1. In form of a calx, *Stannum calciforme.*

    A. Indurated, or vitrified, *Induratum.*

    1. Mixed with a little of the calx of arsenic, *Minera stanni vitrea arsenicalis.*

        a. Solid tin ore, without any determinate figure, Tin-stone.

            It resembles a garnet of a blackish brown colour, but is a great deal heavier; and has been considered at the English tin-mines as a stone, containing no metal, until some years ago it began to be smelted to great advantage.

        b. Cristallised, *Crystallisatum,* Tin-grains.

            Is like the garnets, of a spherical polygonal figure, but looks more unctuous on the surface.

N 3   1. In

1. In larger grains; and,
2. In fmaller grains.

SECT. CLXXXII.

2. Tin mixed with the calx of iron.
   Sect. lxx.
3. Tin mixed with the manganefe.
   Sect. cxvii.
4. Tin mineralifed with fulphur and
   iron, black lead. Sect. cliv.

SECT. CLXXXIII.

OBSERVATIONS on TIN.

It has indeed been afferted by fome, that Tin is found native in the earth; but, for my own part, like many others, I doubt much of it, having never feen a fingle fpecimen that could be called native tin. It is, however, remarkable that tin is fo fcarce, and is not found in any confiderable quantity or purity in any other places in Europe than in England and Saxony. It is likewife worthy obfervation, that when its ore is profitable, or to be worked to any advantage, it is always in form of an indurated calx, which anfwers to thofe glaffes that are prepared from metallic calces in our laboratories: Therefore, in regard to this refemblance, as well as to what this Mineralogical Effay requires from its readers, I have ufed the term *calx*, in defcribing the metals; by which word is underftood the fame as the chemifts call a *crocus*, or *terra metallorum phlogifto privata*.

The tin muft, however, be mineralifed with fulphur in the black lead; but the queftion is, whether that would have happened if the iron
had

had not been prefent? This compound, or black lead, and many more, in which the iron and tin are united, are not eafily to be examined by the common docimaftical means: However, eafier proceffes may poffibly, fome time or other, be difcovered, and employed for fuch fubftances.

SECT. CLXXXIV.

5. Lead, *Plumbum, Saturnus.* It is

a. Of a blueifh white colour when frefh broke, but foon dulls or fullies in the air.

b. Is very heavy; viz. to water as 11,325 to 1000.

c. Is fofteft next to gold, but has no great tenacity, and is not in the leaft fonorous.

d. It is eafily calcined; and, by a certain art in managing the degrees of the fire, its calx becomes white, yellow, and red.

e. This calx melts eafier than any other metallic calx to a glafs, which becomes of a yellow colour, and femi-tranfparent. This glafs brings other bodies, and the imperfect metals, into fufion with it.

f. It diffolves, 1ft, in the fpirit of nitre; 2dly, in a diluted oil of vitriol, by way of digeftion; 3dly, in the vegetable acid; 4thly, in alcaline folutions; and 5thly, in expreffed oils, both in the form of metal and of calx.

g It gives a fweet tafte to all folutions.

h. It amalgamates with quickfilver.

i. With the fpirit of fea-falt it has the fame effect as filver, whereby is produced a *faturnus corneus.*

k. It does not unite with iron, when it is alone added to it in the fire.

N 4   l. It

l. It works on the cuppel, which fignifies that its glafs enters into certain porous bodies, deftitute of phlogifton, and alcaline falts.

m. It melts in the fire before it is made red-hot, almoft as eafily as the tin.

n. Its calx or glafs may be reduced to its metallic ftate by pot-afhes.

SECT. CLXXXV.

Lead is found,

A. In form of a calx, *Minera plumbi calciformis.*

1. Pure, *Minera plumbi calciformis pura.*

   a. Friable, lead ochre, *Ceruffa nativa,* Native cerufs, is found at Kriftierfberget in Weftmanland, on the furface of the potter's ore.

   b. Indurated, Lead fpar, or fpatofe lead ore, *Spatum plumbi.*

      1. Radiated, or fibrous.

         a. White, from Mendip-Hills, in England.

      2. Criftallifed into a prifmatical figure.

         a. White, from Norrgrufva in Weftmanland.

         b. Yellowifh green, from Zchopau in Saxony.

SECT. CLXXXVI.

2. Mixed, *Minera plumbi calciformis mixta.*

   a. With the calx of arfenic, Arfenic lead-fpar.

      1. Indurated.

         a. White. I have tried fuch an ore from an unknown place in Germany, and found that no metallic lead

lead could be melted from it by means of the blowpipe, as can be done out of other lead fpars; but it muft be performed in a crucible, and then that part of the arfenic which did not fly off in fmoke, during the experiment, was likewife reduced, and found in form of grains difperfed, and forced into the lead. Another ore of this kind, which likewife was not eafily reduced by means of the blowpipe, did always after being melted, and during the cooling, haftily fhoot into polygonal, but moftly hexagonal criftals, with fhining furfaces. Can this criftallifation be owing to falts, which are faid not to act in this manner, but when they are diffolved in water?

b. With a calcareous earth *. See Sect, xxxvii.

---

* The abovementioned lead ores are very rich in lead, and eafy to be tried ; becaufe moft of them, being flowly heated, may be reduced to lead by means of the blowpipe on a piece of charcoal. The calx of the lead in thefe ores has, perhaps, firft been diffolved by fulphur and arfenic, and has afterwards, when thefe two have weathered away or decayed, and parted from it, affumed this form ; in the fame manner as we fee it really happens during the calcination, with rich lead ores, or fuch regules as contain lead. The fame, very likely, is the cafe with other metals ; for which reafon their ores, when they occur in form of a calx, often contain a little fulphur, and more efpecially arfenic.

SECT.

SECT. CLXXXVII.

B. Mineralifed, *Plumbum mineralifatum.*

1. With fulphur alone, *Plumbum fulphure mineralifatum:* The *Bley-Schweiff,* or *Bleyglanz,* of the Germans.

   a. Steel-grained lead ore, from the mines at Hellefors, in the province of Weftmanland.

   b. Radiated, or antimoniated lead ore.

   c. Teffellated, or potters lead ore.

   At Villach in Auftria there is faid to be found a potters lead ore, which contains not the leaft portion of filver.

SECT. CLXXXVIII.

2. With fulphurated filver, *Plumbum argento fulphurato mineralifatum. Galena;* alfo called *Bleyglanz* by the Germans.

   a. Steel-grained, is found in the mines of Salberg and Hellefors, in the province of Weftmanland; and in the Dorotheamine, in the Hartz in Germany.

   b. With fmall fcales, is found at Salberg, and is there particularly called *Blyfchweif.*

   c. Fine-grained, found at Salberg.

   d. Of a fine cubical texture; and,

   e. Of coarfe cubes. Thefe two varieties are found in all the Swedifh filver mines.

   f. Criftallifed, from Giflof in the province of Skone *.

---

* The fteel-grained and fcaly ores are of a dim and dull appearance when they are broke, and their particles have no determined angular figure: They are therefore in Swedifh commonly called *Blyfchweif,* in oppofition to the cubical ores, which are called *Blyglanz.* But, in my opinion, the ores ought

to

SECT. CLXXXIX.

3. With fulphurated iron and filver, *Plum-bum ferro & argento fulphurato mineralifa-tum*, is found,
a. Fine-grained,
b. Fine cubical,
c. Coarfe cubical. Thefe are found at Wefterfilfverberget, in Weftmanland.

When this ore is fcorified, it yields a black flag; whereas the preceding lead ores yield a yellow one, becaufe they do not contain any iron.

SECT. CXC.

4. With fulphurated antimony and filver, *Plumbum antimonio & argento fulphurato mineralifatum.* Antimoniated or radiated lead ore. This has the colour of a *Blyglanz*, but is of a radiated texture.

It is found,
a. Of fine rays or fibres, and,
b. Of coarfe rays or fibres.

And is got in Maklos Schacht and Fierde-Bottn, in the mine of Salberg in Weftmanland. The lead in this ore prevents any ufe being made of the an-

to be denominated and diftinguifhed from one another accord-ing to their metallic contents. No ore ought, by virtue of the moft received notion, to be called *Blyfchweif*, but that which contains only lead and fulphur. The moft part of the ores called *Blyglanz* contain filver, even to twenty-four ounces per cent. of which we have inftances in the mines of Salberg, where it has been obferved that the coarfe cubical lead ores are generally the richeft in filver, contrary to what is commonly taught in books; the reafon of which may perhaps be, that, in making the effays on thefe two ores, the coarfe cubical can be chofen purer or freer from the rock, than the fine cubical ores.

timony

timony to advantage; and the antimony likewise in a great meafure hinders the extracting of the filver.

SECT. CXCI.

OBSERVATIONS on LEAD ORES.

I know of no native lead; and all which has been faid on that fubject is liable to remarkable reftrictions.

Such of the potters ores as do not contain any filver are very fcarce; yet they are often found fo poor in filver, that it does not anfwer the ex-pences of extracting. Thefe, when they are free from mixtures of the rock, are, without any pre-vious fufion, employed to glaze earthen-ware; and a great trade is carried on in the Mediterra-nean with fuch ores, from the lead-mines of Sar-dinia and France.

SECT. CXCII.

6. Copper, *Cuprum, Venus, As.*
   This metal is,
   a. Of a red colour.
   b. The fpecific gravity of the Japan copper is 9000, and of the Swedifh 8784 or 8843, to 1000.
   c. It is pretty foft and tough.
   d. The calx of copper being diffolved by acids becomes green, and by alcalies blue.
   e. It is eafily calcined in the fire into a blackifh blue fubftance, which, when rubbed to a fine powder, is red; when melted together with glafs, it tinges it firft
   reddifh

reddifh brown, and afterwards of a tranfparent green or fea-green colour.

f. It diffolves in all the acids; viz. The acids
of vitriol, fea-falt, nitre, and the vegetable; and likewife in all alcaline folutions.
That it becomes rufty, and tarnifhes in the
air (a confequence of a former folution),
depends very much on fome vitriolic acid
which is left in the copper in the refining
of it. This metal is eafier diffolved when
in form of a calx than in a metallic ftate,
efpecially by the acids of vitriol and fea-
falt, and the vegetable acid.

g. Vitriol of copper is of a deep blue colour,
but the vegetable acid produces with the
copper a green falt, which is verdigrife.

h. It can be precipitated out of the folutions
in a metallic ftate; and this is the origin
of the precipitated copper of the mines,
called Ziment copper.

i. It is not eafily amalgamated with quickfilver; but requires for this purpofe a very
ftrong trituration, or the admixture of the
acid of nitre.

k. It becomes yellow when mixed with zink,
which has a ftrong attraction to it, and
makes brafs, pinchbeck, &c.

l. It is eafily diffolved by lead glafs, which
laft is coloured green by it.

m. When this metal is expofed to the fire, it
gives a green colour to the flame in the
moment it begins to melt, and continues
to do fo afterwards, without lofing any
thing confiderable of its weight.

n. It requires a ftrong degree of heat before
it melts, yet is it a leffer degree than for
iron.

SECT.

SECT. CXCIII.

Copper is found in the earth,

A. Native, or in a metallic ſtate; Virgin or native copper, *Cuprum nativum.*

1. Solid, *Solidum,* is found in the iron mine of Heſslekulla, in the province of Nerike, and at Sunnerſkog, in the province of Smoland; alſo in the Ruſſian Carelia, and in other foreign places.

2. Friable, in form of ſmall, and ſomewhat coherent grains, *Cuprum nativum particulis conglomeratis diſtinctis.* Precipitated or Ziment Copper. It is found at Riddarſhyttan in Weſtmanland, at Fahlun in Dalarne, and in Hungary.

It has been obſerved, that both copper and ſilver glaſs ore, being precipitated from water, become friable and granulated, but that they in time grow ſolid and ductile: whence the diſpute about the diſtinction between native and precipitated copper may ceaſe, the rather as native copper will ſcarcely be found in other places, and in any other kinds of ſtones, than through which the ziment or vitriolic waters have circulated; altho' the fiſſures thro' which it has run may afterwards be filled with a ſtony ſubſtance.

SECT. CXCIV.

B. In form of a calx, *Minera cupri calciformis.*

1. Pure, *Minera cupri calciformis pura.*

   a. Looſe or friable, *Ochra veneris.*

   1. Blue, *Cæruleum montanum.*

Is

Is very feldom found perfectly free from a calcareous fubftance.

2. Green, *Viride montanum.*
Both thefe colours depend on men-ftrua, which often are edulcorated or wafhed away.

3. Red. This is an efflorefcence of the glafs copper ore. It is found in the province of Dal, and at Oftanberg, in the province of Dalarne.

SECT. CXCV.

b. Indurated, *Indurata.* Glafs copper ore.
a. Red, *Minera cupri calciformis pura & indurata, colore rubro.*

This is fometimes as red as feal-ing-wax, and fometimes of a more liver-brown colour. It is found in Sandbacken, at Norberg in Weftman-land, at Ordal in Norway, in Siberia, and in Suabia in Germany.

This ore is always found along with native copper, and feems to have loft its phlogifton by way of effloref-cence, and to be changed into this form. It is likewife found along with the fulphurated copper, and is com-monly, though very improperly, called Glafs copper ore.

SECT. CXCVI.

2. Mixed, *Minera cupri calciformis impura.*
a. Loofe or friable, *Ochra veneris friabilis impura.*

1. Mixed

1. Mixed with a calcareous fubſtance, *Ochra veneris terrâ calcareâ mixta. Cæruleum montanum.* In this ſtate copper blue is moſtly found. It ferments during the ſolution in aqua fortis. See Sect. xxxiv.

2. Mixed with iron. Black. It is the decompoſition of the Fahlun copper ore. Sect. cxcviii. *a.*

b. Indurated, *Minera cupri calciformis impura indurata.*

   1. Mixed with gypſum, or plaſter. Green. Is found at Ordal in Norway, and there called Malachites.

   2. Mixed with quartz. Red. From Sunnerſkog, in the province of Smoland. Sect. liii. *B.*

   3. Mixed with lime. Blue. This is the *lapis armenus,* according to the accounts given of it by authors.

S E C T.   CXCVII.

C. Diffolved and mineraliſed, *Cuprum mineraliſatum.*

   1. With ſulphur alone, *Cuprum ſulphure mineraliſatum.* Grey copper ore. Is improperly alſo called Glaſs copper ore.

   a. Solid, without any certain texture, *Minera cupri ſulphurata ſolida texturâ indeterminatâ.* This is very ſoft, ſo that it can be cut with a knife, almoſt as eaſily as black lead.

   b. Fine cubical, *Minera cupri ſulphurata teſſulis conſtans minoribus.*

    Both theſe varieties are found at Sunnerſkog, in Smoland; where the laſt is
ſometimes

fometimes found decompofed or weathered, and changed into a deep mountain blue. Sect. cxciv.

SECT. CXCVIII.

2. With fulphurated iron, *Minera cupri pyritacea:* Yellow copper ore. Marcafitical copper ore, *Pyrites cupri.*

This is various both in regard to colour, and in regard to the different proportion of each of the contained metals; for inftance:

a. Blackifh grey, inclining a little to yellow, *Pyrites cupri grifeus.*

When decayed or weathered, it is of a black colour; is the richeft of all the varieties of this kind of copper ore, yielding between 50 and 60 per cent. and is found in Spain and Germany.

b. Reddifh yellow, or liver brown, with a blue coat on the furface, *Minera cupri lazurea.*

This ore yields between 40 and 50 per cent. of copper, and is commonly faid to be blue, though it is as red when frefh broke, as a rich copper regulus.

c. Yellowifh green, *Pyrites cupri flavo viridefcens.*

This is the moft common in the north part of Europe; and is, in regard to its texture, found

1. Solid, and of a fhining texture, from Oftanberg, in the province of Dalarne.

2. Steel-grained, of a dim texture, from the fame place, and Falun in Dalarne.

3. Coarfe-grained, is of an uneven and fhining texture. It occurs in moft of

O the

the Swedifh and Norwegian copper mines.

4. Criftallifed marcafitical copper ore.

a. Of long octoëdrical criftals.

This is found at Hevaffwik, in the province of Dal, and in Lovifagrufva, in Weftmanland; notwithftanding its exiftence is denied by Henckel, and his followers.

d. Pale yellow, *Pyrites cupri pallidè flavus.*

This cannot be defcribed but as a marcafite, though an experienced eye will eafily difcover fome difference between them. It is found at Tunaberg, in the province of Sodermanland, and yields 22 per cent. of copper.

e. Liver-coloured.

This is found at Falun, in Dalarne, where it contains copper; though at moft other places, where it occurs, it does not contain any copper, but is only a martial marcafite.

SECT. CXCIX.

3. With fulphurated arfenic and iron, *Cuprum ferro ex arfenico fulphurato mineralifatum.* White copper ore.

It is faid to be found in the Hartz, in Germany, and to refemble an arfenical pyrites; but I have never met with this kind.

However, moft of the pyritical copper ores, as well as the marcafites, contain a little arfenic, though it is in too fmall a quantity to be obfervable.

SECT.

SECT. CC.

4. Diffolved by the vitriolic acid, *Cuprum acido vitrioli folutum* : *Vitriolum Veneris.* See Sect. cxxii.

5. With phlogifton. Copper coal ore. See Sect. clxi.

SECT. CCI.

7. Iron, *Ferrum, Mars.* It is

a. Of a blackifh blue fhining colour.

b. It becomes ductile by repeated heating between coals, and hammering.

c. It is attracted by the loadftone, which is an iron ore ; and the metal itfelf may alfo be rendered magnetical.

d. Its fpecific gravity to water is as 7,645, or 8000 : : 1000.

e. It calcines eafily to a black fcaly calx, which, when pounded, is of a deep red colour.

f. When this calx is melted in great quantity with glafs compofitions, it gives a blackifh brown colour to the glafs ; but in a fmall quantity a greenifh colour, which at laft vanifhes, if forced by a ftrong degree of heat.

g. It is diffolved by all falts, by water, and likewife by their vapours. The calx of iron is diffolved by the fpirit of fea-falt, and by aqua regia.

h. The calx of the diffolved metal becomes yellow, or yellowifh brown ; and in a certain degree of heat, it turns red.

i. The

i. The fame calx, when precipitated from acids, by means of the fixed alcali, is of a greenifh colour; but it becomes blue, when precipitated by means of an alcali united with phlogifton, in which laft circum-ftance the phlogifton unites with the iron : Thefe two precipitates lofe their colour in the fire, and turn brown.

k. The vitriol of iron is green.

l. It is the moft common metal in nature, and at the fame time the moft ufeful in common life ; notwithftanding which, its qualities are perhaps very little known.

S E C T. CCII.

Iron is found,

A. In form of calx, *Minera ferri calciformis pura.*

  1. Pure.

    a. Loofe and friable, *Minera ferri calciformis pura friabilis.* Martial ochre, *Minera ochracea.*

      1. Powdry, *Ochra ferri*, is commonly yellow or red, and is iron which has been diffolved by the vitriolic acid.

      2. Concreted. Bog-ore.

        a. In form of round porous balls.

        b. More folid balls.

        c. In fmall flat pieces, like cakes, or pieces of money.

        d. In fmall grains.

        e. In lumps of an indeterminate figure.

          All thefe are of a blackifh brown, or a light brown colour. They are found in lakes in the province of Smoland ;

Smoland; and in marfhes at Fiell-
ryggen, a chain of rocks which feparates Sweden from Norway.

SECT. CCIII.

b. Indurated, *Minera ferri calciformis pura
indurata.* The bloodftone, *Hæmatites.*

1. Of an iron colour, *Hæmatites cærulefcens.*

This is of a blueifh grey colour; it
is not attracted by the loadftone, yields
a red powder when rubbed, and is
hard.

a. Solid, and of a dim appearance
when broken.

b. Cubical, and of a fhining appearance when broken.

c. Fibrous, is the moft common *Torrften* of Sweden.

d. Scaly; the *Eifenman* of the Germans.

This is for the moft part as if it
were micaceous, though the fcales
go acrofs the ftrata of the ftone.
It is found at Jobfbo, in Norrberne
in Dalarne, and Reka Klitt, in the
province of Helfingland.

1. Black, from Gellebeck, in
Norway.

2. Blueifh grey, from Reka Klitt.

When this is found along with
marcafite, as at Sandfwar, in Norway, it is not only attracted by
the loadftone, but is of itfelf
really a loadftone Sect. ccxi.

O 3   *c.* Criftallifed.

c. Criftallifed.
1. In octoëdrical criftals.
2. In polyëdrical criftals.
3. In a cellular form, from Mofsgruf-
van, at Norberg in Weftmanland.

Thefe varieties are the moft common in Sweden, and are very feldom
blended with marcafite, or any other
heterogeneous fubftance, except their
different beds. It is remarkable, that,
when thefe ores are found along with
marcafite, thofe particles, which have
laid neareft to the marcafite, are attracted by the loadftone, although they
yield a red or reddifh brown powder,
like thofe which are not attracted by
the loadftone: It is likewife worth
obfervation, that they generally contain a little fulphur, if they are in-
bedded in a lime-ftone rock, which
however very feldom happens in Sweden; but I know only one fuch inftance, viz. at Billfio, in Soderberke, in
the province of Dalarne. Sect. ccxiii.

S E C T. CCIV.

2. Blackifh brown bloodftone, *Hæma-*
tites nigrefcens. Kidney ore.

This yields a red or brown powder
when it is rubbed; it is very hard,
and is attracted by the loadftone.
a. Solid, with a glaffy texture, from
Wefterfilfverberget, in the province
of Weftmanland.
b. Radiated.
c. Criftallifed.

1. In

1. In form of cones, from Siberia.
2. In form of concentrick balls, with a facetted furface.
   Thefe are very common in Germany, but very fcarce in Sweden.

SECT. CCV.

3. Red bloodftone, *Hæmatites ruber*. Red kidney ore.
   a. Solid, and dim in its texture, from Wefterfilfverberget, in Weftmanland.
   b. Scaly. The *Eifenman* of the Germans. This is commonly found along with the iron-coloured iron glimmer, (Sect. cciii. 1. *d*.) and fmears the hands.
   c. Criftallifed.
      1. In concentrick balls, with a flat or facetted furface.

SECT. CCVI.

4. Yellow bloodftone, *Hæmatites flavus*.
   a. Solid.
   b. Fibrous, from Lammerhof, in Bohemia *.

SECT. CCVII.

2. Iron in form of calx, mixed with heterogeneous fubftances, *Minera ferri calciformis heterogeneis mixta*.   a. With

---

* The varieties of the colours in the bloodftone are the fame with thofe produced in the calces of iron, made by dry or liquid menftrua, and afterwards expofed to different degrees of heat.

a. With a calcareous earth,   White fpa-
thofe iron ore.   The *Stahlftein* of the
Germans.   See Sect. xxx.

b. With a filiceous earth.   The Martial
Jafper or Sinople.   Sect. lxv.

c. With a garnet earth.   Garnet and cockle or fhirl.   Sect. lxix.

d. With an argillaceous earth,   The bole.
Sect. lxxxvi.

e. With a micaceous earth.   Mica.   Sect,
xcv,

f. With manganefe.   Sect. cxvi.

SECT. CCVIII.

g. With an alcali and phlogifton, *Calx
martialis phlogifto junĉta, et alcali preci-
pitata.* Blue martial earth.   Native Pruf-
fian-like blue.

1. Loofe or powdry, found among the
turf in the levels of the province of
Skone : Alfo in Sax Weiffenfels, and
at Norvlanden in Norway, &c.

SECT. CCIX.

h. With an unknown earth, which hardens
in water, *Calx martis terrâ incognitâ
aquâ indurefcente mixta,* Tarras, Cementum,

1. Loofe or granulated, *Terra Puzzolana,*
from Naples and Civita Vecchia in
Italy,   This is of a reddifh brown
colour, is rich in iron, and is pretty
fufible.

2. Indurated, *Cementum induratum,* from
Cologne.

This

This is of a whitifh yellow colour, contains likewife a great deal of iron, and has the fame quality with the former, to harden foon in water, when mixed with mortar. This quality cannot be owing to the iron alone, but rather to fome particular modification of it, occafioned by fome accidental caufes, becaufe thefe varieties rarely happen at any other places, except where volcanos have been, or are yet in the neighbourhood.

SECT. CCX.

i. Calx of iron, united with another unknown earth, *Ferrum calciforme terrâ quâdam incognitâ intimè mixtum.* The *Tungften* of the Swedes.

This is alfo, though improperly, called White Tingrains. Sect. clxxxi.

This refembles the garnet-ftone, (Sect. lxix.) and the tin-grains ; is nearly as heavy as pure tin ; very refractory in the fire, and exceffively difficult to reduce to metal. Iron has, however, been melted out of it to more than 30 per cent. It is very difficultly diffolved by borax and alcaline falts, but melts very eafily with the microcofmic falt, giving a black flag : And for this reafon, this laft mentioned falt muft be employed in the experiments on this ftone. It is found,

1 Solid and fine-grained.

a. Reddifh or flefh-coloured.

b. Yellow.

b. Yellow, from Baſtnaſgrufva at Riddarſhyttan in the province of Weſtmanland.

2, Spathoſe, and with an unctuous ſurface.

a. White, from Marienberg and Altenberg, in Saxony.

b. Pearl - coloured, from Biſpberg
Klack, in the province of Dalarne *.

SECT. CCXI.

B. Diſſolved or mineraliſed iron, *Ferrum mineraliſatum.*

1. With ſulphur alone.

a. Perfectly ſaturated with ſulphur, *Ferrum
ſulphure ſaturatum.* Marcaſite. See Sulphur, Sect. clii.

b. With very little ſulphur. Black iron
ore. Iron ſtone. *Minera ferri atra.*

This is either attracted by the loadſtone, or is a loadſtone itſelf, attracting
iron; it reſembles iron, and yields a
black powder when rubbed.

1. Magnetic iron ore, *Minera ferri attrac-
toria.* The Loadſtone, *Magnes.*

a. Steel-grained, of a dim texture,
from Hogberget, in the pariſh of

---

* This kind of ſtone is very ſeldom met with, but in ſuch
places where black lead is common in the neighbourhood; and
the hiſtory of the black lead, inſerted in the Memoirs of the
Swediſh Academy of Sciences, has induced me to believe, that
this may contain ſome tin, which merits further examination.
Mr. Cronſtedt has in the ſaid Memoirs communicated his experiments upon this kind of ſtone from Riddarſhyttan, and
Biſpberget in Weſtmanland; as has alſo Mr. Rinman, on a
great number of other martial earths. See the ſaid Memoirs
for the years 1751 and 1754.

Gagnœf

Gagnœf in Dalarne: It is found at that place almoſt to the day, and is of as great ſtrength as any natural loadſtones were ever commonly found.

b. Fine-grained, from Saxony.

c. Coarſe-grained, from Spetalsgrufvan, at Norberg, and Kierrgrufvan, both in the province of Weſtmanland. This loſes very ſoon its magnetical virtue.

d. With coarſe ſcales, found at Sandſwœr in Norway. This is a pyritical *Eiſenman,* and yields a red powder when rubbed. Sect. cciii.

SECT. CCXII.

2. Refractory iron ore, *Minera ferri retractoria.* This in its crude ſtate is attracted by the loadſtone.

a. Giving a black powder when rubbed, *Tritura atra.* Of this kind are,

1. Steel-grained, from Adelfors, in the province of Smoland.

2. Fine-grained, from Dannemora, in the province of Upland.

3. Coarſe-grained, from Kierrgrufvan, in the province of Weſtmanland.

This kind is found in great quantities in all the Swediſh iron mines; and of this moſt part of the fuſible ores conſiſt, becauſe it is commonly found in ſuch kinds of rocks as are very fuſible:

And

And it is as feldom met with in quartz, as the hæmatites is met with in limeftone.

SECT. CCXIII.

b. Rubbing into a red powder, *Tritura rubra.*

Thefe are real hæmatites, that are fo far modified by fulphur or lime, as to be attracted by the loadftone.

1. Steel-grained, found in a deferted mine at Billfio, in the parifh of Soderberke in Dalarne.
2. Fine-grained. Emery. This is imported from the Levant: It is mixed with mica, is ftrongly attracted by the loadftone, and fmells of fulphur when put to the fire.
3. Of large fhining cubes, from Thomiensgrube at Arendal in Norway.
4. Coarfe, fcaly. The *Eifenglimmer* or *Eifenman* from Gellebeck in Norway *.

SECT. CCIV.

2. With arfenic, *Ferrum arfenico mineralifatum.* Called Mifpickel by the Germans, and Plate Mundic in Cornwall.
3. With fulphurated arfenic. (Sect. ccxliii.) Arfenical Pyrites.

* Thefe are very fcarce in Sweden, moft part of the Swedifh bloodftones being pure, as has already (Sect. cciii.) been faid, and form that very profitable ore in Swedifh called *Torrften.*

4. With

4. With vitriolic acid. Martial vitriol. Sect. cxxii.

5. With phlogifton. Martial coal ore. Sect. clxi.

6. With other fulphurated and arfenicated metals. See thefe in their refpective arrangements.

SECT. CCXV.

OBSERVATIONS on IRON.

This metal enters into fo many compofitions, that they cannot all be poffibly enumerated; it muft therefore fuffice to mention only thofe, in which it makes out the predominant part. This metal is found in animals and vegetables; and certain iron ores feem to be of fervice to the vegetable kingdom, as is manifeftly feen on the ground round, and under the heaps of loofe ftones laid up in feparating the ore from the rock, at thofe iron mines, where the ores are mixed with limeftone.

With refpect to œconomical effects, iron is divided into cold-fhort, red-fhort, and tough; and the ores into refractory, fufible, and thofe that do not want any admixture; which depends on accidental circumftances, and the method of working.

Although iron is commonly mixed in the different kinds of earth, yet it cannot be afferted with Becher, that iron may be melted out of every earth, by adding only a phlogifton; fince in that cafe this metal might alfo be got out of Mufcovy glafs, pure quartz, chalk, white tranfparent fluor, &c. which very likely has never yet been done.

Nature has beftowed on Sweden an immenfe ftore of iron ores; fo that whole mountains, in Tornea and Lappmark in Lapland, confift folely

of

of a pure, and a very rich iron ore : Large veins of
the fame ore are likewife found in almoft every
province of that kingdom of fuch a nature, that
few countries can produce better or richer.

The magnetical power, with refpect to its principles and origin, is no better underftood than
electricity, yet fomewhat more with refpect to its
effects. Though both thefe qualities are now confidered as different powers, they may perhaps in
time be regarded as fomething nearer allied to
each other.

The magnetical power is not innate in the iron,
but is collected into it by degrees, which is veri-
fied by experiments; it may be expelled, it may
vanifh and gather again, as it were out of the air,
fince the natural loadftones for the moft part occur
in fmall veins to the day, whilft deeper, only refractory iron ores are found. There is the fame difference between an artificial magnet of Dr. Knight's,
and a bar of fteel, whether of the fame fhape or
not, as between a natural loadftone, and a blackifh
blue iron ore ; whence it is ridiculous to infift with
a certain author, that no iron ore can be attracted
by the loadftone, but what of itfelf contains fome
magnetical virtue.

S E C T. CCXVI.

The Second Order.

Semi-Metals.

There are but feven femi-metals yet difcovered.
viz.

1. Quickfilver, Mercury, *Argentum vivum,*
Mercurius, Hydrargyrum.

This diftinguifhes itfelf from all metals, by
the following qualities.   *a.* Its

a. Its colour is white and shining, a little darker than that of silver.

b. It is fluid in the cold, and divisible by the least force; but, as it only sticks to a few bodies, to which it has an attraction, it is said that it does not wet.

c. It is volatile in the fire.

d. Its weight is next to that of the gold, viz. to water, as 13,593 : : 1000.

e. It attracts the other semi-metals and metals, and unites with them all, except cobalt and nickel, with which it cannot by any means yet known be made to mix. This union is called an amalgamation. This amalgamation, or mixtion of metallic bodies, according to the readiness with which they unite or mix, is in the following progression, viz. gold, silver, lead, tin, zink, bismuth, copper, iron, and the regulus of antimony: But the three latter however do not very readily amalgamate. The iron requires a solution of the vitriol of iron, as a medium to promote the union.

f. It dissolves in the spirit of nitre, out of which it is precipitated by a volatile alcali, and the common salt, in form of a white powder; but if a fixed alcali is used, into a yellow powder or calx.

g. It dissolves in the oil of vitriol by a strong boiling.

h. It is not affected by the acid of common salt, unless it be previously dissolved by other acids; in which case only they unite with one another, and may be sublimed together, the which sublimation is a strong poison.

i. It

i. It unites with fulphur by grinding, and then produces a black powder, called *æthiops mineralis*, which fublimes into a red ſtriated body, called factitious cinnabar.

k. The fulphur is again feparated from the quickſilver, by adding iron or lime, to which the fulphur attaches itſelf, leaving the quickſilver to be diſtilled over in a metallic form ; but if a fixed alcali is added to it, fome part of the quickſilver will remain in the reſiduum, and in that caſe makes a liver of fulphur.

S E C T. CCXVII.

Quickſilver is found,

A. Native, or in a metallic ſtate, *Mercurius nativus, virgineus.*

This is found in the quickſilver mines at Idria in Friuli, or the Lower Auſtria, in clay, or in a black ſlaty lapis ollaris, out of which it runs, either ſpontaneouſly, or by being warmed even in the hands. It has feveral times been found at Herr Sten's Bottn, in the mines of Salberg, in Weſtmanland, and fometimes alſo amalgamated with native ſilver.

S E C T. CCXVIII.

B. Mineraliſed, *Mercurius mineraliſatus.*

1. With fulphur, *Mercurius ſulphure mineraliſatus.* Cinnabar, *Cinnabaris nativa.*

This is of a red colour, and its fpecific gravity to water is as 7500 : : 1000.

a. Loofe or friable cinnabar, *Cinnabaris friabilis*, looks like red ochre.

It

It is found in the duchy of Zweybruck or Deuxponts, in Germany.

b. Indurated, *Minera Mercurii indurata*; Solid cinnabar. Is of a deep red colour, and, with refpect to its texture, is either,

1. Steel-grained, from Siebenburgen;
2. Radiated;
3. Compofed of fmall cubes, or fcaly, from Idria and Hungary; or
4. Criftallifed,

    a. In a cubical form; it is tranf-parent, and deep red as a ruby, from Mufchlanfberg in Zweybruck *.

SECT. CCXIX.

2. Mineralifed with fulphur and copper, *Mercurius cupro fulphurato mineralifatus*. This is blackifh grey, of a glaffy texture, and brittle; crackles and fplits exceffively in the fire; and when the quickfilver and ful-phur are evaporated, the copper is difcover-ed by its common opaque red colour in the glafs of borax, which, when farther forced in the fire, or diluted, beccmes green and tranfparent. It is found at Mufchlansberg in Zweybruck.

SECT. CCXX.

OBSERVATIONS on QUICKSILVER.

The divifibility of quickfilver in the cold, might occafion fome doubt, whether it really deferves to be called a metal, if it had not a right to it from

* It is faid, that there is alfo found in Idria a black cinna-bar, that retains its colour in the fublimation, which feems to indicate an abundant phlogifton in the fulphur; but this re-quires, however, a farther examination.

the

the earlieft times, being then reckoned among the
metals, when even they were named after the
planets, the number of both being thought equal.

The opinion, which has a long time prevailed,
that the quickfilver is a neceffary ingredient, and
conftituent part in all metals, is not fo generally
received now as heretofore; fince thofe proceffes,
which have been advanced as proofs of it, and
which have, however, but feldom been repeated,
do by no means fucceed, at leaft not in all places;
it is rather fuppofed, that by the mercurial earth
the ancients have underftood an earth, which may,
by addition of phlogifton, be reduced in the fire
to a metallic ftate; and this appears to be fo much
the more reafonable, as the quickfilver only attracts
the metals in their fubftances, and not in their
burnt calces.

SECT. CCXXI.

2. Bifmuth, Tinglafs, *Vifmutum, Bifmutum,*
Marcafita officinalis. It is
 a. Of a whitifh yellow colour.
 b. Of a laminated texture, foft under the
 hammer, and neverthelefs very brittle.
 c. Its fpecific gravity to water is, as 9,700 : :
 1000.
 d. It is very fufible, calcines and fcorifies like
 lead, if not rather eafier, and therefore
 it works on the cuppel. It is pretty volatile in the fire.
 e. Its glafs or flag becomes yellowifh brown,
 and has the quality of retaining fome part
 of the gold, if that metal has been melted,
 calcined, and vitrified with it.
 f. It may be mixed with the other metals, except cobalt and zink, making them white
 and brittle. *g.* It

g. It diffolves in aqua-fortis, without imparting to it any colour; but to the aqua-regia it gives a red colour, and may be precipitated out of both thefe folutions with pure water, into a white powder, which is called *Spanifh white.* It is alfo precipitated by the acid of fea-falt, which laft unites with it, and makes the *Vifmutum corneum.*

h. It amalgamates eafily with quickfilver. Other metals are fo far attenuated by the bifmuth, when mixed with it, as to be ftrained or forced along with the quickfilver through fkins or leather.

S E C T. CCXXII.

Bifmuth is found in the earth.

A. Native, *Vifmutum nativum.*

This refembles a regulus of bifmuth, but confifts of fmaller fcales or plates.

1. Superficial, or in crufts.
2. Solid, and compofed of fmall cubes.

This is found in, and with, the cobalt ore, at Schneeberg in Saxony, and other foreign places: Likewife along with the copper ore, at Nyberget, in the parifh of Stora Skedwi, in the province of Dalarne.

S E C T. CCXXIII.

B. In form of calx, *Vifmutum calciforme.*

1. Powdry or friable, *Ochra vifmuti.*

This is of a whitifh yellow colour; it is found in form of an efflorefcence, to the day, at Los, in the province of Helfingland.

It has been cuftomary to give the name of Flowers of Bifmuth to the pale red calx

of

of cobalt, but it is wrong ; becaufe neither the calx of bifmuth, nor its folutions, become red, this being a quality belonging to the cobalt.

I have feen a radiated criftallifation of a metallic appearance, which was found at Schneeberg, and was likewife called Bifmuth Flowers ; but in the fmall trial I was permitted to make on it, it did not difcover the leaft marks of bifmuth, but anfwered rather to zink, if zink may be fuppofed to exift in a native ftate.

SECT. CCXXIV.

C. Mineralifed bifmuth, *Vifmutum mineralifatum.*
This is, with refpect to colour and appearance, like the coarfe teffellated potter's lead ore; but it confifts of very thin fquare plates or flakes, from which it receives a radiated appearance, when broken crofswife.

1. With fulphur, *Vifmutum fulphure mineralifatum.*
    a. With large plates or flakes, from Baftnas at Riddarfhyttan, Bafringe and Stripas in Weftmanland.
    b. With fine or fmall fcales, from Jacobsgrufvan at Riddarfhyttan, and the mines at Los, in the parifh of Farila, in Helfingland.

SECT. CCXXV.

2. With fulphurated iron, *Vifmutum ferro fulphurato mineralifatum.*

a. Of

a. Of coarse, wedge-like scales, from Kongruben, at Gellebeck in Norway *.

SECT. CCXXVI.

OBSERVATIONS on BISMUTH.

Although Mr. Pott has, in a separate treatise on bismuth, shewn, that it is dissolved without giving any colour to the solution, and that it is precipitated with pure water; and, though the mine-master Mr. Brandt has likewise, in the *Acta Upsaliensia* for the year 1735, given an accurate history of the cobalt, we find nevertheless in some new authors such a definition of bismuth, as includes at the same time the principal characters of the cobalt, viz that of giving to glass a blue colour, and to tinge solutions red. This confusion proceeds from the bismuth being commonly found among cobalt ores, and that it cannot be separated from it but by the way of eliquation; during which the cobalt, as being less fusible, remains, and is by the workmen called *Vismut graupe*, or Bismuth grains.

This error is excusable in those who do not pretend to maintain and vindicate their ignorance, it having been the fate of the semi-metals to be but very little examined. If the alchemists had not thought the quicksilver, antimony, and zink, fit for their purposes, we should very likely have still wanted many of those advantages which they af-

---

* This mineralised bismuth ore yields a fine radiated regulus; for which reason it has been ranked among the antimonial ores, by those who have not taken proper care to melt a pure regulus or destitute of sulphur from it; while others, who make no difference between regules and pure metals, have still more positively asserted it to be only an antimonial ore.

ford

ford both in medicine and common life. The bifmuth, it is true, has likewife in its time been in fome favour with adepts ; but it foon loft its credit, and was left to thofe who contented themfelves with lefs profpects than of making gold and the univerfal medicine ; as to pewterers, tin-workers, and other tradefmen, who find their advantage in the fufibility of this femi-metal, and its giving co-lour and hardnefs to tin and lead.

S E C T. CCXXVII.

3. Zink, Speltre, *Zincum.*
　　a. Its colour comes neareft to that of lead, but it does not fo eafily tarnifh.
　　b. It fhews a texture, when it is, broken, as if it were compounded of flat pyramids.
　　c. Its fpecific gravity to water is, as 6,900 or 7000 : : 1000.
　　d. It melts in the fire before it has acquired a glowing heat ; but when it has gained that degree of heat, it burns with a flame of a changeable colour, between blue and yellow ; and if in an open fire, the calx rifes in form of foft white flowers ; but if in a covered veffel, with the addition of fome inflammable, it is diftilled in a me-tallic form ; in which operation, however, part of it is fometimes found vitrified.
　　e. It unites with all the metals, except bif-muth, and makes them volatile. It is however not eafy to unite it with iron with-out the addition of fulphur. It has the ftrongeft attraction to gold and copper, and this laft metal acquires a yellow colour by it ; which has occafioned many experi-
　　　　　　　　　　　　　　　　ments

ments to be made to produce new metallic compofitions.

f. It is diffolved by all the acids; of thefe the vitriolic acid has the ftrongeft attraction to it, yet it does not diffolve it, if it is not previoufly diluted with much water. The abundance of phlogifton in this femi-metal is perhaps the reafon of its ftrong attraction to the vitriolic acid.

g. Quickfilver amalgamates eafier with zink than with copper, by which means it is feparated from compofitions made with copper.

h. It feems to become electrical by friction, and then its fmaller particles are attracted by the loadftone; which effects are not yet perfectly inveftigated; but they may excite philofophers to make farther experiments, in order to difcover whether the electrical power fhews itfelf in the metals, by being attracted by the loadftone, or whether the magnetic power can be exerted on other metals than iron.

SECT. CCXXVIII.

Zink is found,

A. In form of calx, *Zincum calciforme naturale.*

1. Pure, *Minera zinci calciformis pura.*

    a. Indurated, *Indurata.*

    1. Solid.

    2. Criftallifed.

        This is of a whitifh grey colour, and its external appearance is like that of a lead fpar; it cannot be defcribed, but is eafily known by an experienced eye. It looks very like an artificial

glafs

glafs of zink, and is found among other calamines at Namur, and in England.

2. Mixed, *Minera zinci calciformis impura.*
   a. With a martial ochre, *Ochra five calx zinci martialis.*
      1. Half indurated, *Ochra zinci indurata.* Calamine, *Lapis calaminaris.*
         a. Whitifh yellow, from Tarnovitz, in Silefia, England, and Aix-la-Chapelle.
         b. Reddifh brown, Poland and Namur. This feems to be a mouldered or weathered blende.
      b. With a martial clay or bole, from Hollberget in Norberke, in Weftmanland. Sect. lxxxvi. *d.*
      c. With a lead ochre and iron, England.

SECT. CCXXIX.

β. Mineralifed zink, *Zincum mineralifatum.*
   1. With fulphurated iron, *Zincum ferro fulphurato mineralifatum.* Blende, Mocklead, Black jack, Mock ore, *Pfeudogalena* and *Blende* of the Germans.
      a. Mineralifed zink in a metallic form, *Zincum formâ metallicâ fulphuratum.* Zink ore.
         This is of a metallic blueifh grey colour, neither perfectly clear as a potter's ore, nor fo dark as the Swedifh iron ores.
         1. Of a fine cubical or fcaly texture, from China, Kongfberg, and Jarlsberg in Norway.

2. Steel-

2. Steel-grained, from Bowallen and Skienfhyttan, in the parifh of Tuna, in Dalarne.

SECT. CCXXX.

b. In form of calx, *Zincum calciforme cum ferro fulphuratum,* Blende. Mocklead, *Sterile nigrum. Pfeudogalena.* This is found,

1. With coarfe fcales,

   a. Yellow, femi-tranfparent, from Scharffenberg in Mifnia, Schemnitz and Kongfberg.

   b. Greenifh, from Kongfberg.

   c. Black, *Pechblende* or *Pitch Blende* of the Germans, from Salberg and Falun in Sweden, and from Saxony.

   d. Blackifh brown, from Storfallfberget in Tuna in Dalarne.

2. With fine fcales,

   a. White, from Silfberget in the parifh of Rettwik in Dalarne.

   b. Whitifh yellow, from Rettwik.

   c. Reddifh brown, from Salberg, Silfverberget, and Hellefors in Weftmanland.

3. Fine and fparkling; at Goflar called *Braun Bleyertz.*

   a. Dark Brown, from the Rammelsberg in the Hartz, and Salberg in Weftmanland *.

SECT.

---

* The zink, in thefe laft kinds of blendes, is as it were in form of a calx or glafs, fo that they are often tranfparent: On the contrary, in the zink ore, (Sect. ccxxix. *a.*) it feems rather to be in a metallic form, or like moft other metals, mineralifed

SECT. CCXXXI.

OBSERVATIONS on ZINK.

It does not feem juſt to conclude from old coins
and other antiquities, that it is evidently proved
that the making of braſs was known in the moſt
antient times, and that it was their *Æs Corinthia-*

neralifed with fulphur. The fulphur, neverthelefs, exiſts in
the different kinds of blende, equally as in the zink ore ; and
this remarkable difference in their appearance muſt be accounted for from another principle than the quantity of the zink
which they contain ; becaufe the yellow and white blendes arc
often found richer than the zink ores, but the zink ores are however more eafy to melt, and confequently more profitable. Perhaps it is becaufe the blende does not contain a fufficient quantity of the phlogiſton of the fulphur, to prevent the calcination of the zink.

It is no matter whether a calcined blende is called calamine
or not, provided it has fuch properties that it may be employed to the fame purpofes. and with the fame advantage as that
calamine which nature has freed from its fulphur by its weathering or decaying. This may be done with fome kinds of
blende, and Mr. Von Swab has given evident and excellent
proofs of it in Sweden ; infomuch that it would demonſtrate
a want of experience to infiſt that fulphur cannot be expelled
by calcination, without deſtroying the zink itſelf, and that
flowers of zink may be produced from zink ores in a calcining
heat, without addition of any phlogiſton.

Mr. Juſti however avers, that he has found an ore of this
quality, which in his Mineralogy he calls *Zirkſpat* ; but there
is great reafon to doubt if it really contains any zink, until it
is proved whether the author added any phlogiſton during the
calcination, or reduced the zink out of it ; becaufe, although the
flowers of zink may not always be perfeɛtly well calcined, yet
there is no inſtance of a natural zink ore being difcovered,
which by itſelf yields thofe flowers during the calcination : And
it requires, befides, a ſtrong heat to produce thefe flowers from
a perfeɛt calx or glafs of this femi-metal, either natural or artificial, though mixed with a phlogiſton ; for it could not have
been a native zink, fince it refembled a fpar, and fuch a one
very likely is not to be found in nature,

cum

cum which contained copper and zink : But, however, it is not long since this semi-metal was discovered to lie concealed in calamine, and that calamine was its particular ore, and also a body of distinct qualities, prepared by nature, equal to that which is got tolerably pure at the furnaces of Gosllar, or that is imported from China, under the name of *Tutanague*. Mr. Brandt removed a great many doubts about the origin of zink, and the metallic earth of the calamine, by having, in the year 1734, a favourable opportunity of examining the calamines, and different kinds of blendes, from Rettwik in the province of Dalarne. He then proved, in his history of the semi-metals, that blendes and calamines are ores of zink, and that the clear alum-like vitriol, called *Galitzenstein* by the Germans, (Sect. cxxii.) was its vitriol. Soon after, the blueish grey zink ore was discovered by Mr. Von Swab at Bowallen, who in the year 1738 prepared calamine from it, and erected a work for distilling zink at large from it, at Westerwiken in Dalarne ; which manufacture, however, afterwards was laid aside for other intervening business. Thus these first discoverers might perhaps have given Messieurs Pott and Margraff the opportunity to write the history of zink, then more known to the world ; the former in his Treatise *De Pseudogalenâ*, in the year 1741; and the latter in the Memoirs of the Academy of Berlin ; though this notice is by no means intended to prejudice these ingenious gentlemen of the honour they merit, to have of themselves had the same opinion, and purposed the same experiments.

The zink ore from Ramelsberg in the Hartz, is, like most of the lead and copper ores from the same mines, of a very fine-grained texture ; and

it

it is likewise so often equally mixed with the said copper and lead ores, as not to be easily perceived, if one is not previously acquainted with them. It seems, nevertheless, reasonable, that a true mineralist ought rather to suspect the ore called *Braunbleyertz* (Sect. ccxxx.) to be a zink ore, than to suppose this semi-metal to be a product of lead, copper, and iron.

SECT. CCXXXII.

4. Antimony, *Antimonium*. *Stibium*. This semi-metal is

a. Of a white colour almost like silver.

b. Brittle; and in regard to its texture, it consists of shining planes, of greater length than breadth.

c. In the fire it is volatile, and volatilises part of the other metals along with it, except gold and platina. It may, however, in a moderate fire be calcined into a light grey calx, which is pretty refractory in the fire, but melts at last to a glass of a reddish brown colour.

d. It dissolves in spirit of sea-salt and aqua regia, but is only corroded by the spirit of nitre into a white calx; it is precipitated out of the aqua regia by water.

e. It has an emetic quality when its calx, glass, or metal, is dissolved in an acid, except when in the spirit of nitre, which has not this effect.

f. It amalgamates with quicksilver, if the regulus, when fused, is put to it; but the quicksilver ought for this purpose to be covered with warm water: It amalgamates
with

with it likewise, if the regulus of antimony be previously melted with an addition of lime.

SECT. CCXXXIII.

Antimony is found in the earth.

A. Native, *Antimonium nativum, five, Regulus Antimonii nativus.*

This is of a silver colour, and its texture is composed of pretty large shining planes.

This kind was found in Carls Ort, in the mine of Salberg, about the end of the last century; and specimens thereof have been preserved in collections under the name of an arsenical pyrites, until the mine-master Mr. Von Swab discovered its real nature, in a treatise he communicated to the Royal Academy of Sciences at Stockholm, in the year 1748. Among other remarkable observations in this treatise, it is said, first, That this native antimony easily amalgamated with quicksilver; doubtless, because it was imbedded in a limestone; since, according to Mr. Pott's experiments, an artificial regulus of antimony may, by means of lime, be disposed to an amalgamation: Secondly, That when brought in form of a calx, it shot into cristals during the cooling *.

---

* Since native antimony, or, as it is commonly called, regulus of antimony, was never before described, the possibility of its existence has been denied; and when this here mentioned was discovered, somebody published some doubts of the truth of the whole affair, upon no better foundation than that the specimens were very small for making experiments, and that it was uncertain if ev r mineralised antimony had been found in the mine of Salberg: but those reasons are not sufficient to refute experiments, because men of experience

are

SECT. CCXXXIV.

B. Mineralifed Antimony, *Antimonium minerali-*
fatum.

   1. With fulphur, *Antimonium fulphure minera-*
lifatum. Antimonium propriè fic dictum.

This is commonly of a radiated texture,
compofed of long wedge-like flakes or
plates ; it is nearly of a lead colour, and
rough to the touch.

  a. Of coarfe fibres.

  b. Of fmall fibres.

  c. Steel-grained, from Saxony and Hungary.

  d. Criftallifed, from Hungary.

     1. Of a prifmatical, or of a pointed
pyramidal figure, in which laft circumftance the points are concentrical.

I have feen a fpecimen of this, in
which the criftals were covered with
very minute criftals of quartz, except
at the extremities, where there was
always a little hole : This fpecimen
was given for a *flos ferri* fpar.

are always able to make true experiments on fmall pieces of
native metals, nor is there any neceflity that mineralifed metals
fhould always be found along with the native ores of the fame
fpecies ; but this really happens with this antimony in the mine
of Salberg. We ought to be contented with conclufions
drawn from experiments, until the fallacioufnefs of fuch experiments is demonftrated : And it were to be wifhed, that all
pretended difcoveries were fupported by experiments, and an
enumeration of the phenomena which happen in them ; we
fhould then not contradict things, which perhaps may be true,
though, for want of this precaution, they feem fcarce credi-
ble ; as, for inftance, the native tin, lead, and iron, the zink
fpar, and an unknown femi-metal in the mica.

SECT. CCXXXV.

2. With fulphur and arfenic, *Antimonium au-ripigmento mineralifatum.* Red antimony ore, *Antimonium folare.*

This is of a red colour, and has the fame texture with the preceding, though its fibres are not fo coarfe.

a. With fmall fibres.

b. With abrupt broken fibres, from Braunf-dorff in Saxony, and from Hungary.

All antimonial ores are fomewhat ar-fenical, but this is more fo than the pre-ceding kinds.

SECT. CCXXXVI.

C. With fulphurated filver. Plumofe filver ore. Sect. clxxiii.

D. With fulphurated filver, copper and arfenic. Sect. clxxiv.

E. With fulphurated lead. Sect. cxc.

SECT. CCXXXVII.

OBSERVATIONS on ANTIMONY.

By the name of Antimony is commonly under-ftood the crude antimony, (which is compounded of the metallic part and fulphur) as it is melted out of the ore (Sect. ccxxxiv.) ; and by the name of Regulus, the pure femi-metal; although this laft begins now to be better diftinguifhed from the other metals.

The alchemifts have very much employed anti-mony in their experiments ; fome of them chiefly

on

on account that it is found in the Hungarian gold
mines : Yet ftill we know no more of the conftituent parts of this femi-metal than the others,
notwithftanding all that has been wrote on the
fubject. Some fay that its earth is not vitrifiable,
becaufe it is volatile, which is perfectly contrary
to experience : And if volatility is the characterif-
tic of a mercurial earth, the pipe clay from Cologne ought to be of the fame nature. Perhaps it
is better to fay that the calx of antimony is volatile, and is incapable of being reduced into a metallic ftate with phlogifton alone, but may be melted into glafs ; and fuch is its nature, though we
do not know the reafon of it.

SECT. CCXXXVIII.

5. Arfenic, *Arfenicum.* This is
   a. In its metallic form nearly of the fame
   colour as lead, but brittle, and changes
   fooner its fhining colour in the air, firft to
   yellow, and afterwards to black.
   b. It appears laminated in its fractures, or
   where broken.
   c. Is very volatile in the fire, burns with a
   fmall flame, and gives a very difagreeable
   fmell, like garlick.
   d. It is, by reafon of its volatility, very difficult to be reduced, unlefs it is mixed with
   other metals : However, a regulus may
   be got from the white arfenic, if it is
   quickly melted with equal parts of pot-
   afhes and foap ; but this regulus contains
   generally fome cobalt, moft of the white
   arfenic being produced from the cobalt
   ores during their calcination. The white
   arfenic, mixed with a phlogifton, fublimes
   like-

likewife into octoëdral criftals of a metallic appearance, whofe fpecific gravity is 8,308.

e. The calx of arfenic, which always, on account of its volatility, muft be got as a fublimation, is white, and eafily melts to a glafs, whofe fpecific gravity is 5,000. When fulphur is blended in this calx, it becomes of a yellow, orange, or red colour; and according to the degrees of colour is called Orpiment or yellow arfenic, Sandarach, Realgar or red arfenic, and alfo *Rubinus Arfenici.*

f. This calx and glafs are diffoluble in water, and in all liquids ; though not in all with the fame facility. In this circumftance arfenic refembles the falts, for which reafon it alfo might be ranked in that clafs. Sect. cxix.

g. The regulus of arfenic diffolves in fpirit of nitre ; but as it is very difficult to have it perfectly free from other metals, it is yet very little examined in various menftrua.

h. It is poifonous, efpecially in form of a pure calx or glafs : But probably it is lefs dangerous when mixed with fulphur, fince it is proved by experience, that the men at mineral works are not fo much affected by the fmoak of this mixture, as by the fmoak of lead ; and that fome certain nations make ufe of the red arfenic in fmall dofes as a medicine.

i. It unites with all metals, and is likewife much ufed by nature itfelf to diffolve, or, as we term it, to mineralife the metals, to which its volatility, and diffolubility in water, muft greatly contribute. It is likewife moft generally mixed with fulphur.

Q

k. It

k. It abforbs or expels the phlogifton, which
has coloured glaffes, if mixed with them
in the fire.

S E C T. CCXXXIX.

Arfenic is found,

1. Native, *Arfenicum nativum*; called *Scherbencobolt* and *Fliegenftein* by the Germans.

It is of a lead colour when frefh broken,
and may be cut with a knife, like black lead,
but foon blackens in the air. It burns with a
fmall flame, and goes off in fmoak.

A. Solid and teftaceous, *Arfenicum nativum
particulis impalpabilibus teftaceum. Scherbencobolt.*

, This is found in the mines of Saxony,
the Hartz, and Hungary.

B. Scaly, *Particulis micaceis*, from Winorn at
Kongfberg in Norway.

C. Friable and porous, *Friabile et porofum.
Fliegenftein.*

1. With fhining fiffures, *Fiffuris nitentibus*,
from Annaberg in Saxony.

This is by fome called *Spigel Cobolt*,
(*Mincra cobalti fpecularis*) according to
their notions of the affinity of thefe metals to one another. However, there always remains after the volatilifation of
the Scherbencobolt, fome calx, either of
cobalt or bifmuth, and fome filver,
though in too fmall a quantity to deferve
any notice.

S E C T. CCXL.

2. In form of a calx, *Arfenicum calciforme.*

A. Pure, or free from heterogeneous fubftances, *Calx arfenici nativa pura.*

1. Loofe

1. Loofe or powdry. This fort is found at Giefshubel in Saxony, but is collected in a much purer ftate on the fides of the rock in fome mines.

2. Indurated or hardened. This is found in form of white femi-tranfparent criftals, in fmall cavities within the Scherbenco-bolt, at Andreafberg in the Hartz, and in Saxony, but is very fcarce.

SECT. CCXLI.

B. Mixed with fulphur, *Calx arfenici fulphure mixta.*

1. Hardened.
   a. Yellow. Orpiment, *Auripigmentum,* from Hungary.
   b. Red. Native Realgar or Sandarach, from Hungary, Andreafberg in the Hartz, Saxony, and Rotendal in Elf-dalen in Sweden *.

SECT. CCXLII.

C. Mixed with the calx of tin, in the tin-grains. Sect. clxxxi.

D. With fulphur and filver, in the Rothgul-den, or red filver ore. Sect. clxx.

E. With calx of lead, in the lead-fpar. Sect. clxxxvi.

F. With calx of cobalt, in the efflorefcence of cobalt. Sect. ccxlviii.

---

* The orpiment may perhaps be found naturally in loofe fcaly powder, as it is fometimes met with in the fhops: However, I have only feen the hardened fort in collections.

SECT. CCXLIII.

3. Mineralifed arfenic, *Arfenicum mineralifatum.*
  A. With fulphur and iron, *Arfenicum ferro fulphurato mineralifatum.* Arfenical pyrites or marcafite *.

This alone produces red arfenic, when calcined, and is found in great quantities in the mines of Lofas in the province of Dalarne : It is of a deeper colour than the following.

  B. With iron only, *Arfenicum metalliforme ferro mixtum. Mifspickel.* This differs with regard to its particles, being

  1. Steel-grained;
  2. Coarfe-grained, from Wefterfilfverberget;
  3. Criftallifed.
    a. In an octoëdral figure. This is the moft common kind.
    b. Prifmatical, from the mines of Salberg, Wefterfilfverberget, and Hellefors in Weftmanland, and in many places of foreign countries †.

SECT. CCXLIV.

  C. With cobalt, almoft in all cobalt ores. Sect. ccxlviii.
  D. With filver. Sect. clxxi. clxxii.
  E. With copper. Sect. cxcix.
  F. With antimony. Sect. ccxxxv.

* Thefe kinds in Cornwall are called Silvery or White Mundics, and Plate Mundics. D. C.

† The fulphureous marcafite is added to this kind, when red arfenic is to be made; but in Sweden it is fcarcer than the fulphureous arfenical pyrites.

SECT.

SECT. CCXLV.

Observations on Arsenic.

Such ores as confift of arfenic united folely with iron, or with iron and fulphur, cannot be employed to any other ufe than to the preparation of arfenical products; for which reafon they ought to be ranged among the arfenic ores. Some have indeed denied this difference between the arfenical pyritæ; but it is however neceffary to make fome difference, with refpect to the prefence or abfence of fulphur, although the greateft quantity of arfenic is got from the calcination of the cobalt ores, and that the true arfenical pyritæ do not deferve to be feparately employed.

Although it is difficult to reduce the arfenic by way of precipitation, one cannot for that reafon deny it to be of a metallic nature; for the fame way of reafoning might have been ufed againft the zink in the calamine, before the method to extract that femi-metal in its metallic ftate, now known, was difcovered: But thofe who know that metals only can be mixed with metals, fo as to preferve the folidity and fome ductility in the compound, and who at the fame time are ignorant of any metallic earth, which cannot be reduced to its metallic ftate again, could never entertain fuch notions.

It is indeed true, that fulphur, in regard to the brittlenefs which it produces in metals, is of no worfe effect than arfenic; but this laft may by itfelf, and mixed only with a pure phlogifton, be fublimed into a metallic form, which is more plainly feen in the Scherbencobolt (Sect. ccxxxix.). I eafily perceive that it may be objected by thofe,

who

who deny arfenic to be a femi-metal, that it may
as well be a falt of a peculiar nature; as, for inftance, the vitriolic acid; and that it may, like fulphur, diffolve the metals in form of a kind of regulus; and farther, that its affuming a metallic
appearance, when it is united with an inflammable fubftance, is of no confequence, fince there
are fifh and infects who have a fhining metallic colour: But all this does not deferve an anfwer,
fince it has been already agreed, that fyftems muft
not be too feverely criticifed.

S E C T. CCXLVI.

6. Cobalt, *Cobaltum.*
  This femi-metal is,
  a. Of a whitifh grey colour, nearly as fine
     tempered fteel.
  b. Is hard and brittle, and of a fine grained
     texture; hence it is of a dufky, or not fhin-
     ing appearance.
  c. Its fpecific gravity to water is 6000 :: 1000.
  d. It is fixt in the fire, and becomes black by
     calcination; it then gives to glaffes a blue
     colour, inclining a little to violet, which
     colour, of all others, is the moft fixed in
     fire.
  e. The concentrated oil of vitriol, aqua fortis,
     and aqua regia, diffolve it; and the folu-
     tions become red. The cobalt calx is like-
     wife diffolved by the fame menftrua, and
     alfo by the volatile alcali, and the fpirit of
     fea-falt.
  f. When united with the calx of arfenic in a
     flow (not a brifk) calcining heat, it affumes
     a red colour: The fame colour is naturally
     produced by way of efflorefcence, and is
                                        then

then called the *bloom* or *flowers of cobalt*. When cobalt and arfenic are melted together in an open fire, they produce a blue flame.

g. It does not amalgamate with quickfilver by any means hitherto known.

b. Nor does it mix with bifmuth, when melted with it, without addition of fome medium to promote their union.

SECT. CCXLVII.

The cobalt is moft commonly found in the earth mixed with iron.

A. In form of a calx, *Cobaltum calciforme.*

1. With iron without arfenic, *Martiale abfque arfenico.*

   a. Loofe or friable, *Minera cobalti calciformis pulverulenta.* Cobalt ochre, *Ochra cobalti nigra.* It is black, and like the artificial zaffre.

   b. Indurated, *Minera cobalti calciformis indurata. Minera cobalti vitrea,* the *fchlacken* or flag cobalt.

   This is likewife of a black colour, but of a glaffy texture, and feems to have loft that fubftance which mineralifed it, by being decayed or weathered. It is often confounded with the Scherbenco-bolt, for it is feldom quite free from arfenic; and there may perhaps exift a progreffive feries from the Schlacken kind to the Scherbencobolt kind.

SECT. CCXLVIII.

2. With the calx of arfenic, *Minera cobalti calciformis calce arfenici mixta*. Cobalt-blut, *Ochra cobalti rubra*, Bloom, Flowers, or Efflorefcence of cobalt.

   a. Loofe or friable, *Ochra cobalti pulverulenta*. This is often found of a red colour like other earths, fpread very thin on the cobalt ores, and is, when of a pale colour, erroneoufly called Flowers of Bifmuth.

   b. Indurated, *Ochra cobalti rubra indurata*. Hardened Flowers of Cobalt.

     This is commonly criftallifed in form of deep red femi-tranfparent rays or radiations: It is found at Schneeberg in Saxony *.

SECT. CCXLIX.

B. Mineralifed, *Cobaltum mineralifatum*.

   1. With arfenic and iron in a metallic form, *Cobaltum ferro & arfenico metalliformis mineralifatum*; *vulgò* Cobaltum *dictum*.

    This is of a dim colour when broken, and not unlike fteel. It is found

   a. Steel-grained, from Loos in the parifh of Farila, in the province of Helfingeland, and at Schneeberg in Saxony.

   b. Fine-grained, from Loos,

---

* A white cobalt-earth, or ochre, is faid to have been found. It has been feen and examined by a celebrated mineralift, who has found it in every refpect, except the colour, to refemble the cobalt flowers; and it is very poffible, that thofe cobalt flowers might in length of time have loft their red colour, and become white.

                      c. Coarfe-

c. Coarfe-grained.
d. Criftallifed.
  1. In a dendritical or arborefcent form, from Schneeberg.
  2. Polyëdral, with fhining furfaces; the *Glanzkobolt* of the Germans, from Schneeberg.
  3. In radiated nodules, from Kongfberg in Norway.

S E C T. CCL.

2. With fulphurated iron, *Cobaltum ferro fulphurato mineralifatum.*

This is of a lighter colour than the preceding, nearly like to tin or filver. It is found

a. Criftallifed.
  1. In a polygonal form.
    a. Of a flaggy texture.
    b. Coarfe-grained.

    This kind is found in Baftnasgrufva át Riddarfhyttan in Weftmanland, and difcovers not the leaft mark of arfenic. The coarfegrained becomes flimy in the fire, and fticks to the ftirring hook during the calcination, in the fame manner as many regules do; and is a kind of regule prepared by nature.

    That fort of a flaggy texture is very martial, and is defcribed by the mine-mafter Mr. Brandt, in the Acts of the Swédifh Academy of Sciences for the year 1746. Both thefe give a beautiful colour.

S E C T.

SECT. CCLI.

3. With fulphur, arfenic, and iron, *Cobaltum cum ferro fulphurato et arfenicato mineralifatum.*

This refembles the arfenicated cobalt ore, being only rather of a whiter or lighter colour. It is found

a. Coarfe-grained.

b. Criftallifed.

    1. In a polygonal figure, with fhining furfaces, or *Glanzkobolt.*

        It occurs at Tunaberg in the province of Sodermanland, partly of a white or light colour, and partly of a fomewhat reddifh yellow.

SECT. CCLII.

4. With fulphurated and arfenicated nickel and iron; fee *Kupfernickel,* Sect. cclvi.

SECT. CCLIII.

OBSERVATIONS on COBALT.

Since the glafs of cobalt, which has been entirely freed from all arfenic in the calcination, and from the iron and the other metals by fcorification, as when it is prepared from criftallifed cobalt flowers, may by addition of phlogifton be melted to a true cobalt regulus, which differs in its qualities from all other metals; there can be no reafon for denying the cobalt a place among the femi-metals, as many authors even at this time do, notwithftanding

withftanding the feveral reafons given, which might induce them to examine nearer into the fubject.

It was the mine-mafter Mr. Brandt who firft difcovered this femi-metal, and defcribed it in the abovementioned Hiftory of Semi-metals, in the *Acta Upfalienfia* for the year 1735.

The brittlenefs of the cobalt regulus is no proof againft its being a femi-metal, that property being the bafis on which the diftinction between the femi-metals and metals is founded. The earth of cobalt is fixed and vitrifiable in the fire, as well as that of copper and iron; and the colour of its glafs being fo immutable in the fire, proves it to be a particular fubftance, diftinct from other earths and metallic calces. The experiment of making a cobalt glafs from iron or fteel and arfenic, will certainly never fucceed, unlefs the arfenic, employed for that purpofe, has been made from a cobalt ore; but if the origin of the colour fhould be afcribed to an irreducible metallic earth, there is no occafion for this experiment, becaufe a cobalt regulus may be prepared fo as to be free both from arfenic and iron, the prefence of this laft metal being eafily difcovered by the loadftone. It is therefore now unneceffary and ridiculous to continue the old definitions of the cobalt, in which the Speife, which partly is a cobalt regulus, and partly a compound, confifting of nickel, cobalt, and bifmuth, united with fulphur and arfenic, is either confounded with the femi-metal itfelf, or quoted as a proof, that a cobalt regulus cannot exift in any other manner than as a dead earth involved in heterogeneous fubftances; which is the fame as to conclude, that no pure copper can be produced from the copper regules or fufions, called *Trottften* or *Spurften*.

These

These false notions have, however, induced a new author to describe the cobalt as a mixture of iron, copper, lead, bismuth, and arsenic; but he has not at the same time published any experiments which might serve to confirm his opinion; amongst which, with great reason, such experiments are expected as imitate nature in this composition, which is pretended to consist of so many different things. It might then have been calculated, if it would be profitable to establish manufactures for making cobalt-glass, or zaffre, in any part of the world, where the above-mentioned ingredients can be had.

The word Cobalt in Germany, and especially at the mineral works in Saxony, is applied to the damps, the arsenic, its vapours, and their effects on man; which has induced the vulgar also to apply it to some pretended evil spirit, which is said to dwell in the mines: But time will abolish these superstitions, which have their origin from ignorance.

SECT. CCLIV.

7. Nickel, *Niccolum*.

This is the latest discovered semi-metal. It was first described by its discoverer Mr. Cronstedt in the Acts of the Royal Academy of Sciences at Stockholm, for the years 1751 and 1754, where it is said to have the following qualities: That

1. It is of a white colour, which however inclines somewhat to red.

2. Of a solid texture, and shining in its fractures.

3. Its specific gravity to water is as 8,500 :: 1000.

4. It

4. It is pretty fixt in the fire; but together with the fulphur and arfenic, with which its ore abounds, it is fo far volatile, as to rife in form of hairs and branches, if in the calcination it is left without being ftirred.

5. It calcines to a green calx:

6. This calx is not very fufible, but however tinges glafs of a tranfparent reddifh brown, or jacinth colour.

7. It diffolves in aqua fortis, aqua regia, and the fpirit of fea-falt, but more difficultly in the vitriolic acid, tinging all thefe folutions of a deep green colour. Its vitriol is of the fame colour; but the colcothar of this vitriol, as well as the precipitates from the folutions, become by calcination of a light green colour.

8. Thefe precipitates are diffolved by the fpirit of fal ammoniac, and the folution has a blue colour; but being evaporated, and the fediment reduced, there is no copper, but a nickel regulus is produced.

9. It has a ftrong attraction to fulphur; fo that when its calx is mixed with it, and puts on a fcorifying teft under the muffel, it forms with the fulphur a regule: This regule refembles the yellow fteel-grained copper-ores, and is hard and fhining on its convex furface.

10. It unites with all the metals, except quickfilver and filver. When the nickel regulus is melted with the latter, it only adheres clofe to it, both the metals lying near one another on the fame plane; but they are eafily feparated with a hammer. Cobalt has the ftrongeft attraction to nickel,

after

after that to iron, and then to arfenic. The two former cannot be feparated from one another but by their fcorification, which is eafily done, fince

11. This femi-metal retains its phlogifton a long time in the fire, and its calx is re- duced by the help of a very fmall portion of inflammable matter: It requires, how- ever, a red heat before it can be brought into fufion, and melts a little fooner, or almoft as foon as copper or gold, confe- quently fooner than iron.

S E C T.   CCLV.

The nickel is found.

A. In form of a calx, *Niccolum calciforme.* Nickel ochre, *Ochra niccoli.*

1. Mixed with the calx of iron, *Ochra niccoli martialis.*

This is green, and is found in form of flowers on Kupfernickel. In Normarken in the province of Wermeland, this ochre was found without any vifible nickel mixed in the clay, which contained a great quan- tity of native filver. Sect. clxviii.

S E C T.   CCLVI.

B. Mineralifed nickel, *Niccolum mineralifatum.*

1. With fulphurated and arfenicated iron and cobalt, *Niccolum ferro & cobalto arfenicatis et fulphuratis mineralifatum. Cuprum nicolai feu niccoli. Kupfernickel.*

This is of a reddifh yellow colour, and is found

a. Of a flaggy texture, in Saxony.

b. Fine-

b. Fine-grained, and

c. Scaly, in Loos cobalt mines in the province of Helfingeland; at which place it is of a lighter colour than the foreign ones. Thefe two are often from their colour confounded with the liver-coloured marcafite. Sect. cliii.

SECT. CCLVII.

2. With the acid of vitriol, *Niccolum acido vitrioli mineralifatum.*

This is of a beautiful green colour, and may be extracted out of the nickel ochre, (Sect. cclv.) or efflorefcence of the Kupfernickel.

SECT. CCLVIII.

OBSERVATIONS on NICKEL.

The cobalt, bifmuth, and nickel, are commonly found together in the fame mines, by which it happens, that, when the firft, as the moft ufeful of them all, is to be made into glafs, the adherent nickel, according to its nature, unites with the fulphur and arfenic, of which fome portion remains after the calcination, and makes with them a regule. When thefe minerals (the fulphur and arfenic) are in greater quantity than is wanted for the nickel, they likewife reduce fome part of the calces of the cobalt and bifmuth; and in this cafe the nickel, as a medium, uniting the other two, otherwife not mifcible femi-metals, incorporates them into the fame regule. From hence arifes a difference in the contents of different regules; and from this difference people, who
have

have not fufficient experience, form to themfelves
falfe notions of the whole compound, and of each
part contained in it: For which reafon they chufe
rather to retain that definition of the Kupfernickel
which has received its fanction from the earlieft
authors, than to admit the conclufion to which
Mr. Cronftedt's experiments feem to lead.

For my own part, I have found myfelf obliged to
follow the opinion of the latter, partly becaufe I
am tired with thofe common epithets given to unknown bodies; fuch as, *wild, refractory, rapacious,
arfenical, irreducible, metallic earth*, &c. &c. which
regard the effect alone and not its caufe; and partly becaufe I have not, befides the nickel, found
any metal or metallic compofition, which

1. Becomes green when calcined.

2. Yields a vitriol, whofe colcothar alfo becomes green in the fire.

3. So eafily unites with fulphur, and forms
with it a regule of fuch a peculiar nature, as the
nickel does in this circumftance; and that

4. Does not unite with filver, but only adheres
or fticks clofe to it, when they have been melted
together.

The nickel not having yet been found free from
cobalt and iron, is the reafon why it was not difcovered.   This was the cafe alfo with the cobalt.
Platina del pinto perhaps, in the fame manner, might
for a long time have been mixed in the gold, at
certain places, where it is faid to be naturally paler
than any where elfe in the world. But the exiftence
of fuch things cannot any longer be denied, fince
the method is difcovered to get them feparate,
and free from heterogeneous fubftances. It indeed
would be the fame thing, as if in a country where
filver is never found but in the potter's lead ore,
any perfon fhould deny the exiftence of either of
thefe

thefe metals, or infift upon it, that one is produced from the other.

It is remarkable, that the precipitates of nickel give a blue colour to the fpirit of fal ammoniac, when they are diffolved in it; without fhewing befides any marks of copper, which, however, could not be concealed if there were any; for if a fmall quantity of copper is melted with the nickel, and kept in a ftrong fire with it, the copper foon feparates, and fcorifies, tinging the glafs firft of a reddifh brown opaque colour, and, the fire being further forced, it then makes it tranfparent and green, as ufual.

There is no danger attending the encreafing the number of the metals. Aftrological influences are now in no repute among the learned, and we have already more metals than planets within our folar fyftem. It would perhaps be more ufeful to difcover more of thefe metals, than idly to lofe our time in repeating the numberlefs experiments which have been made, in order to difcover the conftituent parts of the metals already known. In this perfuafion, I have avoided to mention any hypothefes about the principles of the metals, the proceffes of mercurification, and other things of the like nature, with which, to tell the truth, I have never troubled myfelf.

APPENDIX.

SECT. CCLIX.

I Have already in the Preface mentioned the reafons why the Saxa and Foffils commonly called Petrefactions, cannot be ranked in a mineral fyftem: And I am almoft perfuaded, that the fame reafon which has prevailed on me, will likewife after mature confideration be approved by others. Mean while, fince thefe bodies, efpecially the latter, occupy fo confiderable a place in moft mineral collections, and the former muft neceffarily be taken notice of by the miners in the obfervations they make in the fubterranean geography, I would not entirely omit them here, but have tried to put them in fuch an order as may anfwer that purpofe, for which miners and mineralifts pay any regard to them.

SECT. CCLX.

The FIRST ORDER.

SAXA. PETRÆ.

I divide thefe into two kinds.

1. Compound Saxa, *Saxa compofita*,

Are fuch ftones whofe particles, confifting, of different fubftances, are fo exactly fitted and joined together, that no empty fpace, or even cement, can be perceived between them; which feems to indicate, that fome, if not all,

of

of thefe fubftances have been foft at the inftant of their union.

2. Conglutinated ftones, *Saxa conglutinata*,

Are fuch ftones whofe particles have been united by fome cementitious fubftance, which, however, is feldom perceivable, and which often has not been fufficient to fill every fpace between the particles: In this cafe the particles feem to have been hard, worn off, and in loofe, fingle, unfigured pieces, before they were united.

S E C T. CCLXI.

1. Compound Saxa, *Saxa compofita*.

A. Ophites. Scaly limeftone with kernels or bits of ferpentine ftone in it, *Saxum compofitum particulis calcareis et argillaceis*.

1. *Kolmord's marble.* It is white and green.

2. *Serpentino antico*, is white, with round pieces of black fteatites in it. This muft not be confounded with the *Serpentino verde antico*. Sect. eclxvi.

3. The *Haraldfio marble.* White, with quadrangular pieces of a black fteatites.

4. The *Marmor Pozzevera di Genoua.* Dark green marble, with white veins.

This kind receives its fine polifh and appearance from the ferpentine ftone.

S E C T. CCLXII.

B. Stellften or *Geftellftein*, *Saxum compofitum particulis quartzofis & micaceis*.

1. Of diftinct particles, *Particulis diftinctis*.

This is found at Garpenberg in the province of Dalarne. It is likewife met

R 2   with

with in the other mineral mountains of Sweden. In fome of thefe the quartzofe particles predominate, and in others, the micaceous : In the laft cafe it is commonly flaty, and eafy to fplit.

2. Of particles which are wrapt up in one another, *Particulis quartzofis micâ convolutis.*

a. Whitifh grey, from Morthernberget in Norberke in Dalarne.

b. Greenifh, at Salberg in Weftmanland.

c. Reddifh, from the parifh of Malung in Dalarne.

Both thefe kinds of Stellften are for their refiftance to the fire employed in building furnaces; but the latter is the beft, becaufe it feems at the fame time to contain a little of a refractory clayifh fubftance : It however cracks very foon, if the flat fide of the ftratum, inftead of the extremity, is turned towards the fire. It is alfo of great ufe in mills, if the other or fellow-ftone is made of the mill-ftone from Arfunde, which is a Saxum of the conglutinated kind, or a coarfe fand-ftone. It is lucky for œconomical purpofes, that the plates of thefe ftones are fo thick, although thereby they are not fo eafily fplit.

SECT. CCLXIII.

C. Norrka. Murkften of the Swedes. *Saxum compofitum micâ, quartzo et granato.*

1. With diftinct garnets or fhirl, *Granatis diftinctis cryftallifatis.* *a.* Light

a. Light grey, from Selbo in Norway.

b. Dark grey, with very fmall garnets, from Quarnberget in the parifh of Soderli in the province of Jemteland.

c. Dark grey, with prifmatical, radiated, or fibrous cockle or fhirl, from the village of Handol, in the parifh of Are in Jemteland.

2. With kernels of garnet ftone, *Particulis granatinis indeterminatis.*

a. Of pale red garnet ftone, from Stollberget in Norberke in Dalarne.

The firft of this kind, whofe flaty ftrata makes it commonly eafy to be fplit, is employed for mill-ftones, which without difficulty diftinguifh themfelves for that purpofe, if fand is firft ground with them, becaufe the fand wears away the micaceous particles on the furfaces, and leaves the garnets prominent, which renders the ftone fitter for grinding the corn.

S E C T. CCLXIV.

D. The Whetftone, *Cos. Saxum compofitum micâ, quartzo, et forfan argillâ martiali in nonnullis fpeciebus.*

1. Of coarfe particles, *Particulis diftinctis.*

a. White, from Wanga in the province of Skone.

b. Light grey, from Tellemarken in Norway.

2. Of fine particles, *Particulis minoribus.*

a. Liver brown colour, from Selbo in Norway.

R 3 *b.* Blackifh

b. Blackifh grey, from Lerwik at Hellefors in Weftmanland, and from Cologne in Germany.

c. Light grey, from Hellefors in Weftmanland.

d. Black. The table flate, or that kind
ufed for large tables and for fchool-
flates.

The naked eye, and the magnifying glafs much better, difcovers the
micaceous particles in this kind to be
as it were twifted in one another; fome
clay feems likewife to enter into the
compofition: However, it cannot yet
be certainly afferted that it is real
mica which has that appearance in
this kind.

3. Of very minute and clofely combined
particles, *Cos particulis conftans impalpabilibus durus.* The Turky ftone.

This is of an olive colour, and feems
to be the fineft mixture of the firft fpecies of this genus. It is found in loofe
ftones at Biorkfkoginas in the parifh of
Hellefors in Weftmanland, though not
perfectly free from crofs veins of quartz,
which always are in the furface of the
rock, and fpoil the whetftones. It is
alfo faid to be found in Tellemarken in
Norway. The beft of this fort come
from the Levant, and are pretty dear.
The whetftone kinds, when they fplit
eafily, and in thin plates, are very fit to
cover houfes with, though moft of them
are not ufed for that purpofe.

SECT.

SECT. CCLXV.

E. The *Telgſten* of the Swedes.   *Lapis olla-*
ris. Saxum compoſitum ſteatite et micâ.

a. Light grey, from Fahlun, and alſo
Byxberget at Norberke.

b. Whitiſh yellow, from Sikſioberget in
Norberke.

c. Dark grey, from Riddarſhyttan.

d. Dark green, from Salviſto in the pariſh
of Tamela in Finland.

This is employed with great advantage to build fire-places and furnaces,
&c. and when it is ſlaty, the extremities of the ſtrata muſt be turned towards
the fire.

SECT. CCLXVI.

F. Porphyry, *Porphyrites.* Italorum *Porfido.*
Saxum compoſitum jaſpide et feltſpato, inter-
dum micâ et baſalte.

a. Its colour is green, with light green
feltſpat, *Serpentino verde antico.* It is
ſaid to have been brought from Egypt
to Rome, from which latter place the
ſpecimens of it now come.

b. Deep red, with white feltſpat, from
Italy, and Egern in Norway.

c. Black, with white and red feltſpat, from
Klitten, in the pariſh of Elfdalen in Dalarne.

d. Reddiſh brown, with light-red and
white feltſpat, from Hykieberget in Elf-
dalen, and Guſtavſtrom, in the pariſh of
Goſborn in the province of Wermeland.

R 4   e. Dark

e. Dark grey, with white grains of feltſpat alſo, from Guſtavsſtrom.

Many varieties of this kind, in regard to colour, are found in form of nodules or looſe ſtones in Sweden; but I have only mentioned the hardeſt and fineſt of thoſe which are found in the rocks; becauſe, beſides theſe, there are coarſe porphyries found, which ſcarce admit of any poliſh. The dark red porphyry has been moſt employed for ornaments in building: yet it is not the only one known by the name of *Perſido*, the Italians applying the ſame name alſo to the black kind.

SECT. CCLXVII.

G. The *Trapp* of the Swedes. *Saxum compoſitum jaſpide martiali molli, ſeu argillâ martiali induratâ, et - - . - -*

This kind of ſtone ſometimes conſtitutes or forms whole mountains; as, for example, the mountain called Hunneberg, in the province of Weſtergottland, and at Drammen in Norway; but it is oftener found in form of veins in mountains of another kind, running commonly in a ſerpentine manner, contrary or acroſs to the direction of the rock itſelf. It is not homogeneous, as may be plainly ſeen at thoſe places where it is not preſſed cloſe together; but where it is preſſed cloſe, it ſeems to be perfectly free from heterogeneous ſubſtances. When this kind is very coarſe, it is interſperſed with feltſpat; but it is not known if the finer ſorts likewiſe contain

any

any of it. Besides this, there are also some fibrous particles in it, and something that resembles a calcareous spar: This however does not ferment with acids, but melts as easy as the stone itself, which becomes a black solid glass in the fire. By calcination it becomes red, and yields in assays 12 or more per cent. of iron. No other sort of ore is to be found in it, unless now and then somewhat merely superficial lies in its fissures; for this stone is commonly, even to a great depth in the rock, cracked in acute angles, or in form of large rhomboidal dice. It is employed at the glass-houses, and added to the composition of which bottles are made. By the Germans it is called *Schwach* or *Schwartsstein*; at the Swedish glass-works, *Trappskiol*, *Tegelskiol*, or *Swartskiol*; and at Jarllberg in Norway, *Blabest*. In the air it decays a little, leaving a powder of a brown colour; it cracks commonly in the fire, and becomes reddish brown if made red hot. It is found

1. Of coarse chaffy particles, *Particulis majoribus acerosis:*
    a. Dark grey, from the top of Kinnekulle in the province of Westergottland.
    b. Black, from Stallberget at Osterfilfverberget in the province of Dalarne.
2. Coarse-grained, *Particulis majoribus granulatis.*
    a. Dark grey, from the uppermost stratum at Hunneberg in Westergottland.
    b. Reddish, from Bragnas in Norway.
    c. Deep brown, from Gello in Norway.

3. Of

3. Of fine imperceptible particles, *Particulis impalpabilibus.*

    a. Black. The Touchſtone, *Lapis Lydius,* from Salberg mine, Hellefors, Weſterſilfverberget, and Norberg in Weſtmanland, and Oſterſilfverberget in Dalarne, &c.

    b. Blueiſh, from Oſterſilfverberget.

    c. Grey, from Dalwik in the pariſh of Sorberke in Dalarne.

    d. Reddiſh, from Dalſtugun in the pariſh of Rettwik in Dalarne *.

S E C T. CCLXVIII.

H. Amygdaloides. Saxum baſi jaſpideâ martiali, cum fragmentis ſpati calcarei et ſerpentini, figurâ ellipticâ †.

It is a martial jaſper, in which elliptical kernels of calcareous ſpar and ſerpentine-ſtone are included.

    a. Red, with kernels of white limeſtone, and of a green ſteatites, from Gello and Gullo in Norway, and the Hartz in Germany.

This is of a particular appearance, and when calcined is attraƈted by the loadſtone ; it decays pretty much in the air, and has ſome affinity with the Trapp, (Seƈt. cclxvii.) and alſo with the Porphyry (Seƈt. cclxvi.). There are ſome-

---

* The black variety (3. *a.*) is ſometimes found ſo compaƈt and hard, as to take a poliſh like the black agat; it melts, however, in the fire to a black glaſs, and is, when calcined, attraƈted by the loadſtone. Such a kind is found in the pariſh of Arla in the province of Sodermanland.

† The Carpolithi or Fruit-ſtone rocks of the Germans. D. C.

times

times found pieces of native copper in this ftone at Gullo.

SECT. CCLXIX.

J. The *Gronften* of the Swedes. *Saxum compoſitum micâ et hornblende.* Sect. lxxxviii.

Its bafis is hornblende, interfperfed with mica. It is of a dark green colour, and is dug in feveral places in Smoland, where it is employed in the iron furnaces as a flux to the bog ore (Sect. ccii.). It is alfo found in other places, as at Rettwik in Dalarne, and in the neighbourhood of fome of the iron mines.

SECT. CCLXX.

K. The Granites. *Saxum compofitum feltfpato micâ et quartzo, quibus accidentaliter interdum hornblende, fteatites, granatus et bafaltes immixti funt.*

Its principal conftituent parts are feltfpat, or rhombic quartz, mica, and quartz: It is found

1. Loofe or friable, *Particulis conftans parum coherentibus.*

This is ufed at the brafs works to caft the brafs in, and comes from France.

2. Hard and compact, *Granites durus.*
    a. Red.
        1. Fine-grained, from Swappawari at Tornea in Lapland
        2. Coarfe-grained, from Bifpbergs Klack, in the province of Dalarne.
    b. Grey, with many and various colours, found on the coaft round Stockholm and Norland,   The

The Granites are feldom flaty or laminated, when their texture is clofe, and the harder particles, as the feltfpat or rhombic quartz, the quartz, and the fhirl, predominate in it. They admit of a good polifh, for which reafon the Egyptians in former times, and the Italians now, work them into large pieces of ornamental architecture, for which purpofe they are extremely fit, as they do not decay in the air.

SECT. CCLXXI.

2. Conglutinated Saxa, *Saxa conglutinata.*

A. Of larger or broken pieces of ftones of the fame kinds conglutinated together, *Saxum conglutinatum fragmentis lapidum. Breccia.*

1. Of limeftone cemented by lime, *Saxum conftans fragmentis lapidis calcarei, calce conglutinatis.*

a. The calcareous Breccia, *Breccia calcarea:* The *Marmi Brecciati* of the Italians.

When thefe kinds have fine colours, they are polifhed and employed for ornaments in architecture, and other œconomical ufes; they come from Italy.

b. The *Lumachella* of the Italians, or fhell marbles. Thefe are a compound of fhells and corals, which are petrified or changed into lime, and conglutinated with a calcareous fubftance. When they have many colours, they are called

led Marbles, and employed for the fame purpofes as the preceding; likewife from Italy, from Bergen in Norway, and Offerdal in the province of Jemteland. In the ifland of Gottland there is found of this kind of one colour only, which on that accôunt is not called Marble, or ufed as fuch. At Balsberget, in the province of Skone, is found a white and yellow fhell limeftone, of weak colours.

S E C T. CCLXXII.

2. Of kernels of jafper cemented by a jafpry fubftance, *Saxum fragmentis jafpidis materiâ jafpideâ conglutinatum. Breccia Jafpidea. Diafpro brecciato* of the Italians.

Of this kind fpecimens from Italy are feen in collections. A coarfe Jafper Breccia is faid to be found not far from Frejus in Provence in France.

S E C T. CCLXXIII.

3. Of filiceous pebbles, cemented by a jafpry fubftance, or fomething like it, *Saxum filicibus amorphis materiâ jafpideâ conglutinatis.* The plum-pudding ftone of the Englifh. *Breccia Silicea.*

Its bafis, which at the fame time is the cement, is yellow, wherein are contained fingle flinty or agaty pebbles, of a grey colour or variegated. This is of a very elegant appearance when cut and polifhed; it is found in England.

S E C T.

SECT. CCLXXIV.

4. Of quartzofe kernels combined with an unknown cement, *Saxum fragmentis quartzofis conglutinatis*. *Breccia quartzofa*. Found in the provinces of Jemteland and Smoland.

SECT. CCLXXV.

5. Of kernels of feveral different kinds of ftones, *Saxum fragmentis variorum faxorum conglutinatis*. *Breccia faxofa*.

a. Of kernels of porphyry, cemented by a porphyry or coarfe jafpry fubftance, *Breccia porphyrea*, from Serna Fiell, and Hykieberget in the province of Dalarne.

b. Of kernels of feveral faxa, *Saxum fragmentis variorum faxorum compofitorum conglutinatis*. *Breccia indeterminata*.

Is found in loofe ftones in Dalarne, and are originally broken from the Fiell tracts in Serna, which confift of nothing elfe but conglutinated ftones.

c. Of conglutinated kernels of fandftone, *Saxum fragmentis conftans faxorum conglutinatorum*. *Breccia arenacea*.

This kind confifts of fandftone (Sect. cclxxvi.) kernels, which have been combined a fecond time together. This is alfo found in loofe ftones in Dalarne, and are perhaps originally broken

broken from the above-mentioned
Fiell in Serna *.

S E C T. CCLXXVI.

B. Conglutinated ftones of granules or fands
of different kinds, *Saxum conglutinatum granulis feu arenâ variorum lapidum.* Sandftone,
Lapis arenaceus.

In this divifion are reckoned thofe which
confift of fuch minute particles, that all
of them cannot eafily be difcovered by the
naked eye. The greateft part, however,
confift of quartz and mica, which fubftances are the moft fit to be granulated,
without being brought to a powder.

I think I have reafon to confider this kind
in regard to the fubftance which has ferved
as a cement to combine them, although it
is not always perfectly difcernible.

1. Cemented by clay, *Lapis arenaceus glutine argillaceo.*

* The above-m ntioned Brecciæ of themfelves muft demand
the diftinctions here made between them, but which perhaps
may feem to be carried too far, fince their particles are fo
big and plain, as to be eafily known from one another.
Thefe ftones are a proof both of the fubverfions which the
mountains in many centuries have undergone, and of fome
hidden means which Nature makes ufe of in thus cementing
different kinds of ftones together. Any certain bignefs for the
kernels or lumps in fuch compounds, before they deferve the
name of Breccia, cannot be determined, becaufe that depends
on a comparifon, which every one is at liberty to imagine.
At one place in the mountain called Hykieberget, the kernels
of porphyry have a diameter of fix feet, while in other places
they are not bigger than walnuts. At Maffewala the kernels
have a progreffive fize down to that of a fine fandftone. Moft
of this kind of ftone is fit for ornaments, though the workmanfhip is very difficult and coftly.

a. With

a. With an apyrus or refractory clay, *Argillâ porcellaneâ.*

It is found under the stratum of coal in the coal-mine at Boserup in Skone; is of a loose texture, but hardens, and is very refractory in the fire.

b. With common clay, *Argillâ communi,* from Burswick in the island of Gottland.

2. With lime, *Lapis arenaceus glutine calcareo*; resembles mortar made with coarse sand.

a. Consisting of transparent and greenish grains of quartz and white limestone; from the island Ifo, near Beckaskog in Skone.

b. Of no visible particles, from France and Livonia. This is of a loose texture, and hardens in the air.

3. With an unknown cement, *Lapis arenaceus glutine incognito, forsan argillaceo.*

a. Loose, from Helsingberg in Skone.

b. Harder, from Roslagen, Orsa, and Kinnekulle.

c. Compact, from Gefle in the province of Gestrikeland, and the lake Malaren.

d. Very hard, from Serna Fiell or Fells in Dalarne; it is also found in great abundance in loose stones at Gustavsstrom, and at Siliamfors in the parish of Mora in Dalarne.

4. Cemented by the rust or ochre of iron, *Lapis arenaceus ochrâ martis conglutinatus.* Is found in form of loose stones at several places, and ought perhaps to be reckoned among

among the *Mineræ Arenaceæ* or Sand-
Ores; (Sect. cclxxvii.) at leaft when the
martial ochre makes any confiderable
portion of the whole *.

SECT. CCLXXVII.

C. Stones and ores cemented together, *Saxum
fragmentis conftans lapidum et minerarum conglutinatis. Mineræ Arenaceæ.*
  1. Of larger fragments, *Fragmentis lapidum et minerarum majoribus.*
    a. Mountain green, or *Viride Montanum
    Cupri*, and pebbles cemented together,
    from Siberia.

* Sandftones are of great ufe in œconomy, as materials for
buildings which refift fire, air, and water, are made of them.
Some of them are foft while in the quarry, but harden in the
open air. The loofe fandftones are the moft ufeful; but the
folid and hard ones, fuch as *c.* and *d.* crack in the fire, and
take a polifh when ufed as grindftones. However, though the
Burfwik's ftone, (1. *b.*) is pretty loofe in its texture, it is neverthelefs unfit for buildings which are expofed to fire or open air,
becaufe it breaks to pieces and melts in the fire, and in the air
it attracts the moifture, decays in length of time, and cracks
in the cold, which proceeds from the included kernels of clay
expanding themfelves when they grow wet; the fandftones
ought therefore to be very nicely examined before they are
employed to the ufual purpofes.
There are many quarries of fandftones in Sweden, but no
enquiry has yet been made if any of them, and which, can be
employed in the larger works, inftead of the Englifh, and in
the fmaller manufactories inftead of the Bohemian fandftones.
Such enquiries are of greater confequence, in proportion as
thofe manufactures encreafe wherein they are wanted. It muft
be remarked, that the working-mafons or ftone-cutters ought
to wear a piece of frieze or baize before their mouth and nofe,
in order to preferve themfelves from a premature death, which
now-a-days unhappily is the cafe with them in the parifh of
Orfa in Dalarne, and in other places; but limeftone is not
found to have this bad effect.

S *b.* Potters

b. Potters lead ore, with limeſtone, ſlate-kernels and ſhells, from Gragruf-van at Boda in Rettwik, and in Da-larne.

c. Yellow or marcaſitical copper ore, with ſmall pebbles.

S E C T. CCLXXVIII.

2. Of ſmaller pieces, *Granulis lapidum et minerarum.*

 a. Potters lead ore with a quartzoſe ſand, from Eiffelsfeldt near Cologn in Germany.

 b. Mountain green with ſand, from Si-beria.

 c. Cobalt ore with ſand.

 d. Martial ochre with ſand *.

S E C T. CCLXXIX.

OBSERVATIONS on the SAXA or STONES.

Beſides the advantages which may accrue to œconomy by a perfect knowledge of the Saxa, the miners or ſubterranean geographers expect alſo

---

* The *Mineræ Arenaceæ* or Sand-ores, cannot reaſonably be ſeparated from the ſand-ſtones, ſince they are produced in the ſame manner ; beſides, when they are poor in yield, they are alſo employed to the ſame purpoſe, becauſe it is not eaſy to ſmelt the metal out of them. The ſand-ores, beſides, cannot be ranked in a mineral ſyſtem as ſeparate ſpecies of ores, be-cauſe they would then be arranged with reſpect to the kind of ſtone in which the ore occurs, and not the ore itſelf, which caſe cannot be admitted here. It might be urged, that ores, mixed with the ſtones of the very load, and not in form of ſand-ores, ought as well as them to be ranked among the com-pound ſaxa ; but in that caſe there would be no end of ſpecies, nor could they ever be reduced into any order.

another

another future benefit from it, viz. that of con-
cluding, from many obfervations, if all the Saxa
are to be equally confidered; for example, if in
fome of them veins or ftrata of ores may be expeƈt-
ed, and if thofe are only of certain kinds ; if others
are every where found deftitute of any ore whatfoever ; if, and which of them are fit to form coats
on the furface of the rock, which covers other
kind of ftones, and alfo veins and ftrata of ores,
&c. If no general rules are to be deduced from
fuch obfervations, there is a probability, at leaft, to
gain fome inlights that may be particular to certain
countries; and this opinion is already in fome
places confirmed by experience. Hence it may
be concluded, how neceffary it is to communicate
all fuch obfervations which, for the above-mentioned purpofes, ought to be made over the whole
globe, and to agree on fixing certain names on the
Saxa, in order to avoid too great a prolixity in their
defcriptions. It is with this intention I have here,
as a trial, given fpecific names to thofe Saxa which
are found in this northern country, and which Saxa
I know; wifhing at the fame time to be acquainted with a method to diftinguifh them more eafily
and to better purpofe.

This procedure will be found ftill more neceffary
and ufeful, as the world feems refolved foon to
abolifh the fuperftition of the Hazel Rod or *Vir-
gula Divinatoria*, and that we have by means of
obfervations already got too much experience to
believe, that the ftrata of earths and ftones are
placed equally and in the fame order and fituation
over the whole earth; which fome, however, in
thefe our times have even endeavoured to prove,
while others have made a fecret of it, in order by
fome way or other to enrich themfelves.

SECT. CCLXXX.

The SECOND ORDER.

MINERAL-CHANGES, or the PETREFACTIONS.

Mineralia-Larvata, vulgò *Petrefacta*,.

Are mineral bodies in the form of animals or vegetables, and for this reason no others belong to this order, than such as have been really changed from the subjects of the other two kingdoms of nature.

There is more difficulty to determine the first point, viz. from when these bodies are to be stiled petrefactions, than from when they cease to be such; meanwhile I have, in order to make a trial, considered them in the following manner.

SECT. CCLXXXI.

1. Earthy Changes, *Terræ Larvatæ. Terrificata.*.
   A. Extraneous bodies changed into a lime substance, or calcareous changes, *Larvæ calcareæ.*
   1. Loose or friable. Chalky changes, *Cretæ larvatæ.*
      a. In form of vegetables.
      b. In form of animals.
         1. Calcined or mouldered shells, *Humus conchaceus*, from the province of Helsingeland, at Uddevalla in the province of Halland, and in the French strata of earth and chalk.
   2. Indurated, *Petrefacta calcarea.*
      a. Changed

a. Changed and filled with folid limeftone.

,1. In form of animals.

2. In form of vegetables.

Found in the ifland of Gottland.

b. Changed into a calcareous fpar, *Petrefaƈta calcarea fpatofa.*

1. In form of animals.

The fhells in Balfberget in the
province of Skone.

2. In form of vegetables *.

S E C T. CCLXXXII.

B. Extraneous bodies changed into a flinty
fubftance. Siliceous changes, *Larvæ filicææ.*
Thefe are like the flint,

1. Indurated, *Petrefaƈta filicea.*

a. Changed into flints.

* Shells and corals are indeed compofed of limy matter,
even when their animals ftill dwell in them ; neverthelefs, although they are not changed in regard to their principle, yet
are they reckoned among the petrefaƈtions, as foon as the
particles of the calcareous fubftance have got a new fituation ;
for example, when they are become fparry, when they have
been filled with a calcareous earth, either hardened or loofe,
or when they lie in the ftrata of the earth. Thefe form
the greateft part of foffil colleƈtions, which are fo induftrioufly
made, often without any regard to the only and principal ufe
they can be of, viz. that of enriching zoology. Mineralifts
are fatisfied with feeing the poffibility of the changes the limeftone undergoes in regard to its particles, and alfo with re-
ceiving fome inlight into the alterations which the earth has
been fubjeƈt to, from the ftrata which are now found in it.
The calcined fhells, or thofe which have been changed into
limy and chalky matter, are fit to make lime, and are ftill
more ferviceable as a manure. The indurated ferve only to
make grottos. No gypfeous petrefaƈtions are known, if fuch
are not found in the Perfian alabafter ; for Mr. Chardin fays,
that he has feen a lizard included in that ftone.

1. Carnelians

1. Carnelians in form of shells, from the river Tomm in Siberia.
2. Agat in form of wood. Such a piece is said to be in the collection of Count Teffin.
3. Coralloids of white flint, (*Millepora*) found in Gottland.
4 Wood of yellow flint. Italy, Adrianople, and Loughneagh, a lake in Ireland.

S E C T. CCLXXXIII.

C. Extraneous bodies changed into clay. Argillaceous changes, *Larvæ argillaceæ.*
a. Loose and friable.
  1. Of porcellane clay.
    a. In form of vegetables.

A piece of white porcellane clay from Japan, with all the marks of the root of a tree, has been observed in a certain collection.

    b. Indurated.
      1. In an unknown clay.
        a. In form of vegetables. *Osteocolla.* It is said to be changed roots of the poplar tree, and not to confist of any calcareous fubstance. See the *Physicaliche Beluftigungen.*

A fort of foffil ivory is said to be found, which has the properties of a clay; but I do not know if it is rightly examined,

SECT. CCLXXXIV.

2. Saline extraneous bodies, or such as are penetrated by mineral salts, *Corpora peregrina insalita. Larvæ insalitæ.*
A. With the vitriol of iron, *Vitriolo martis insalita.*
   1. Animals.
     a. Human bodies have been twice found in the mine at Falun in Dalarne; the last was kept a good many years in a glafs cafe, but began at laft to moulder and fall to pieces.
   2. Vegetables.
     a. Turf, and
     b. Roots of trees.
     Thefe are found in water ftrongly impregnated with vitriol; for inftance, in the moor at Ofterfilfberget in Dalarne. They do not burn with a fiame, but only like a coal in a ftrong fire; neither do they decay in the air.

SECT. CCLXXXV.

3. Extraneous bodies penetrated by mineral inflammable fubftances, or mineral phlogifton, *Corpora peregrina phlogifiis mineralibus impregnata.*
A. Penetrated by the fubftance of pit-coals, *Lithantrace impregnata.*
   1. Vegetables, which commonly have been woods, or appertaining to them.
     a. Fully faturated. *Gagas. Jet.*

The

The jet is of a folid fhining texture. From England, Boferup in Skone, and the Black Sea.

b. Not perfectly faturated. *Mumia vegetabilis.* Is loofe, refembles umbre, and may be ufed as fuch. From Boferup.

SECT. CCLXXXVI.

B. Penetrated by rock oil or afphaltum, *Corpora peregrina petroleo feu afphalto impregnata.*
1. Vegetables.
   a. Turf, in the province of Skone.
   The Egyptian mummies cannot have any place here, fince art alone is the occafion that thofe human bodies have in length of time been penetrated by the afphaltum, in the fame manner as has happened naturally to the wood in pit-coal ftrata (Sect. cclxxxv. *b*).

SECT. CCLXXXVII.

C. Penetrated by fulphur which has diffolved iron, or by marcafite and pyrites, *Pyrite impregnata. Petrefacta pyritacea.*
1. Human.
   a. Bivalves,
   b. Univalves, and
   c. Infects.
   In the alum flate at Andrarum in Skone.

SECT.

SECT. CCLXXXVIII.

4. Metals in the form of extraneous bodies,
Larvæ metalliferæ.
A. Silver, *Larvæ argentiferæ.*
  1. Native.
     a. On the furfaces of fhells. England.
  2. Mineralifed with copper and fulphur.
     a. Fahlertz or grey filver ore (Sect. clxxi.) in form of ears of corn, &c. and fuppofed to be vegetables, are found in argillaceous flate at Frankenberg and Tahlitteren in Heffe.

SECT. CCLXXXIX.

B. Copper, *Larvæ cupriferæ.*
  1. Copper in form of calx, *Cuprum calciforme corpora peregrina ingreffum.*
     a. In form of animals, or of parts belonging to them.
        1. Ivory, and other bones of the elephant. The Turcois or Turky ftone. It is of a blueifh green colour, and much valued in the Eaft.

        At Simore in Languedoc bones of animals are dug, which during the calcination affume a blue colour; but it is not probable that the blue colour is owing to copper.

SECT. CCXC.

2. Mineralifed copper, which impregnates extraneous bodies, *Cuprum mineralifatum corpora peregrina ingreffum.*

A. With

A. With fulphur and iron. The yellow or marcafitical copper ore that
impregnates

1. Animals.

 a. Shells, from Hagatienns Schurff
and Jarlfberg in Norway. Thefe
fhells lye upon a loadftone,

 b. In form of fifh, from Eeifleben,
Mansfeld, and Otterode, in Germany.

B. With fulphur and filver. Grey filver
ore or Fahlertz, like ears of corn, from
the flate quarries in Heffe (Sect.
cclxxxviii.).

SECT. CCXCI.

C. Changes into iron, *Larvæ ferriferæ.*

1. Iron in form of calx, which has
affumed the place or the fhape of
extraneous bodies, *Ferrum calciforme
corpora peregrina ingreffum.*

 a. Loofe, *Larvæ ochraceæ.*

  1. Of vegetables.

   Roots of trees, from the
lake Langelma in Finland:
See the Acts of the Swedifh
Academy of Sciences for the
year 1742.

 b. Indurated, *Larvæ hæmatiticæ.*

  1. Of vegetables.

   Wood, from Orbiffau in
Bohemia.

SECT. CCXCII.

2. Iron mineralifed, affuming the
fhape of extraneous bodies, *Ferrum
mineralifatum*

mineralifatum corpora peregrina in-
greffum.
   a. Mineralifed with fulphur. Marcafite. *Larvæ Pyritaceæ.* Sect.
   cclxxxvii.

SECT. CCXCIII.

5. Extraneous bodies decompofing, or in a way
of deftruction, *Corpora peregrina in gradibus
deftructionis confiderata.* Mould, *Humus.* Turf,
Turba.
A. From animals. Animal mould, *Humus
   animalis.*
   1. Shells. *Humus conchaceus.*
   2. Mould of other animals, *Humus diver-
     forum animalium.*
B. Vegetable mould, *Humus vegetabilis.*
   1. Turf, *Turba.*
     a. Solid, and hardening in the air, *Turba
       folida aëre indurefcens.* Is the beft of
       this kind to be ufed for fuel, and
       comes neareft to the pit-coals. It
       often contains a little of the vitriolic
       acid.
     b. Lamellated turf, *Turba foliata.* This
       is in the firft degree of deftruction.
   2. Mould of lakes, *Humus lacuftris.* This
     is a black mould which is edulcorated
     by water.
   3. Black mould, *Humus ater.*
     This is univerfally known, and covers
     the furface of that loofe earth in which
     vegetables thrive beft *.

---

* All the kinds of mould contain fome of the inflammable
fubftance, which has remained in them from the vegetables or
animals; and they are more or lefs black, in proportion as they
                        contain

SECT. CCXCIV.

The THIRD ORDER.

NATURAL SLAGS,

Scoriæ Vulcanorum.

Slags are found in great abundance in many places in the world, not only where volcanos yet exift, but likewife where no fubterraneous fire is now known: Yet, according to our opinion, they cannot be produced but by means of fire. Thefe are not properly to be called natural, fince they have marks of violence, and of the laft change that mineral bodies can fuffer without the deftruction of the world; nor are they artificial, according to the univerfally received meaning of this word. When we perhaps in future times by new difcovered means may be able to find out of what fort of earth ftones are compounded, we fhall ftill be forced to ftop at the furface of them, and be contented with knowing that they contain a little iron. Mean while I cannot omit them here, fince I have confidered the petrefactions; and therefore I will enumerate fome of them, according to their external marks.

contain more or lefs of this phlogifton. I have ranked them in this place, that they might not be totally excluded. They are elfe a *medium uniens* between all the three kingdoms of nature: And it may reafonably be afked, if all forts of earth do not in form of very minute particles enter into the compofition of vegetables and animals, after which they exift for fome time in form of mould, until the phlogifton is again feparated.

SECT.

SECT. CCXCV.

1. Iceland agat, *Achates iſlandicus niger.*
It is black, ſolid, and of a glaſſy texture; but in thin pieces, it is greeniſh and ſemi-tranſparent like glaſs bottles, which contain much iron. The moſt remarkable is, that ſuch large ſolid maſſes are found of it, that there is no poſſibility of producing the like in any glaſshouſe.
It is found in Iceland, and in the Iſland of Aſcenſion: The jewellers employ it as an agat, though it is too ſoft to reſiſt wear.

SECT. CCXCVI.

B. Rheniſh millſtone, *Lapis molaris Rhenanus.*
Is blackiſh grey, porous, and perfectly reſembles a ſort of ſlag produced by Mount Veſuvius. If I am miſtaken in this, I hope that ſomebody elſe will deſcribe the conſtituent parts of this millſtone.

SECT. CCXCVII.

C. Pumice-ſtone, *Pumex.*
Is very porous and bliſtered, in conſequence of which it is ſpecifically very light. It reſembles that froathy ſlag which is produced in our iron furnaces.
1. White.
2. Black.
The colour of the firſt is perhaps faded or bleached, becauſe the ſecond kind comes in that ſtate from the laboratory itſelf, viz. the volcanos.

SECT.

S.E C T. CCXCVIII.

D. Pearl flag, *Scoriæ conſtantes globulis vitreis conglomeratis.*

Is compounded of white and greeniſh glaſs particles, which ſeem to have been conglutinated while yet ſoft, or in fuſion, Found on the Iſle of Aſcenſion.

S E C T. CCXCIX.

E. Slag-ſand or aſhes, *Scoriæ pulverulentæ. Cineres Vulcanorum.*

This is thrown forth of the volcanos in form of larger or ſmaller grains. It may perhaps be the principle of the Terra Puzzolana (Sect. ccix. *a.*), becauſe ſuch an earth is ſaid at this time to cover the ruins of Herculaneum near Naples, which hiſtory informs us was deſtroyed by a volcano during an earthquake.

S E C T. CCC.

OBSERVATIONS on the preceding SLAGS.

It ſeems as if we could not go any farther in the arrangement of bodies belonging to the mineral kingdom, than to the black mould (Sect. ccxciii.) and the flags, as being the extremes.

However, if theſe flags likewiſe decay, and in length of time become an earth, which poſſibly may happen; there is then a new ſubſtance beyond them, which however may return back and circulate again in ſome known form. It is obvious how the old heaps of flags from the iron

furnaces

furnaces decay, and at laft produce vegetables, which cannot be afcribed to a black mould alone carried thither by the wind. The fame may perhaps happen with the natural flags in the open air; but we do not know if it is fo, nor what different forms this and every other earth which circulates in animals and vegetables further affumes: However, in fuch circumftances, as their particles become or are already very minute, and moft part of the phlogifton becomes volatile, when acted upon by heat or fire, it feems probable, that, by a flow feparation of the phlogifton, or a union by means of falts, this earth is moft apt to become a clay, provided it is not by any previous revolution laid in fuch places as to change it into flate, pit-coal, &c.

If at any time it fhould happen that a volcano fhould burft out of a mountain, whofe ftrata we knew before, we could at leaft imagine fome reafons for this wonderful effect: However, the learned would neverthelefs, perhaps, want fome knowledge about the fubftances of the ftrata, and the manner of their formation; fince in this circumftance water and other obftacles have hindered people too much from making the due obfervations thereon.

Meanwhile, the more we confider, on the one part, all the modifications and alterations the earths undergo by means of fire and water, by the free or impeded accefs of the air, by the volatility and attraction of the acid falts, whereby are produced folution and hardening, compofition and feparation; and, on the other part, reflect on the fhortnefs of a man's life, perhaps alfo dedicated to other bufinefs, on the difficulty of obferving the fubterraneous effects, and on feveral other things, which prevent the making difcoveries, by which we might find out fome eafier means to attain true

knowledge

knowledge by judicious experiments; the more we shall find, what is wanted to form mineral fyftems, and for this reafon be apt to excufe the faults of thofe which have been hitherto publifhed.

From thofe, who of themfelves are fufceptible of thefe fentiments, I fuffer with pleafure that judgment, which I am myfelf ready to pronounce upon this Effay,

Tranfeat cum ceteris.

F I N I S.

DESCRIP.

DESCRIPTION and USE

OF A

Mineralogical Pocket Laboratory;

AND ESPECIALLY THE

USE of the BLOW-PIPE

IN

MINERALOGY.

By GUSTAV von ENGESTROM.

T

DESCRIPTION

OF A

Mineralogical Pocket Laboratory, &c.

SECT. I.

THAT Science which teaches us the properties of mineral bodies, and by which we
learn how to characterize, diftinguifh, and
clafs them into a proper order, is called *Mineralogy*. This, like all other fciences, when rightly
cultivated, and employed to its proper end, (the
Public Good) furnifhes us with many ufeful difcoveries, in proportion as it increafes.

SECT. II.

Mineralogy has been ftudied for feveral ages,
yet its progrefs has been very flow.

Some learned men have, indeed, endeavoured to
bring it into fome fyftematical order: But as the
paffion for only collecting minerals and foffils has
ftill predominated over that of diving into the
nature of the fubjects themfelves, they have for
the moft part met with but very little fuccefs.
Thofe who were *mere Collectors*, being fuperior in
number to the *fcientifical ones*, or *Mineralifts*, and

having

having more opportunities of getting new specimens, were moſt of them not ſo communicative
to the latter as they ought to have been. Some
of theſe, fond of the number, were wholly taken
up in gathering together immenſe heaps of things,
ſeeming almoſt reſolved to get the whole of Nature
into their cabinets, without having regard to any
true order; while others, purpoſing to correct
this inconveniency, would pretend to ſome interior knowledge, as if that had been a conſequence of their collection; and by that fell into
another ſtill greater extravagancy.

All this certainly hindered the mineraliſts from
improving much in the ſcience; but, happily,
thoſe times are paſt. The world is grown more
reaſonable at preſent, and Mineralogy ſeems more
and more to be encouraged.

The great utility of the mineral bodies already
known, promiſes us a greater advantage from the
ſtudy of this ſcience, than only the pleaſure of collecting. But, in order to come at this advantage,
we ought to ſearch into the very principles of
theſe bodies, that we may be certain of not deceiving ourſelves in our judgment about them.

SECT. III.

As the principal end of cultivating Mineralogy
is to find out the œconomical uſe of the minerals,
it is neceſſary to know every occurrent mineral
body in regard to all its effects; from them to
determine the beſt uſe it might be put to. A
Syſtem of Mineralogy thus founded on the effects
of its ſubjects, muſt be more ſcientifical, ſince it
always has in view that real point, *their application
to Common Life :* And ſince it is natural to the
human mind to adapt every thing to its own advantage,

vantage, as far as poffible, fuch a fyftem muft be
more generally received, and at the fame time
the eafier underftood, as it includes the mineral
bodies in a lefs number of claffes, orders, &c. by
which the memory is not fo much clogged, as if
only their furfaces had been defcribed.

SECT. IV.

This granted, let us confider what difficulties
there are to be met with in examining mineral bodies.
Thefe are often like one another as to their external
appearances, although their conftituent parts are
quite different, and confequently make them ufeful in different ways: Moft part of them ought
alfo to be changed from their natural form, and
even often diffolved, before they can be made any
ufe of. Their figure and colour, or, in fhort, their
furfaces, are therefore not folely to be depended
upon; we muft penetrate into them; and they
muft be decompounded according to the principles of chemiftry.

SECT. V.

By examining the mineral kingdom in this manner, we may now and then find the fubjects of
our experiments (if even nearly the fame) to differ
in fome of their effects, which is particularly owing
to the difficulty of juftly determining the degrees
of the fire employed; a difficulty not yet removed, but which, however, ought not to hinder
us from going as far as poffibly we can, fince
we find by practice, that fuch obftacles often are
remedied by repeated experiments; and of thefe
we never can make too many, if judicioufly performed.

T 3 SECT.

SECT. VI.

This way of studying Mineralogy has already some time ago been entered upon; but Mr. Pott, at Berlin, has brought it to a greater perfection; and after him Mr. Cronstedt, in Sweden, has extended it yet farther, submitting every mineral body, that came to his hands, to chemical experiments; in consequence of which he afterwards published his *Essay towards a System of Mineralogy.*

SECT. VII.

Thus the greatest obstacle is removed; the best method to learn Mineralogy is laid open, in following which we are enabled to render this Science more and more perfect. To obtain this end, chemical experiments are without doubt necessary; but as a great deal of the mineral kingdom has already been examined in this manner, we do not want to repeat all those experiments in their whole extent, unless some new and particular phœnomena should discover themselves in those things we are examining; else the tediousness of those processes might discourage some from going farther, and take up much of the time of others, that might be better employed. An easier way may therefore be made use of, which even for the most part is sufficient, and which though made in miniature, yet is as scientifical as the common manner of proceeding in the laboratories, since it imitates that, and is founded upon the same principles. This consists in *a method of making experiments upon a piece of charcoal with the concentrated flame of a candle blown through a Blow-pipe.* The

heat

fig. 1.

fig. 2. *fig. 3.*

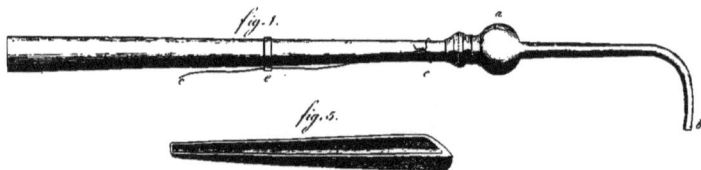

fig. 4.

fig. 5.

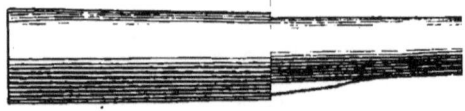

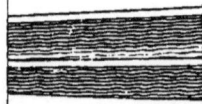

5.

heat occafioned by this, is very intenfe, and the
mineral bodies may here be burnt, calcined, melted, or fcorified, &c. as well as in any great works.

SECT. VIII.

The Blow-pipe is in common ufe among jewellers, goldfmiths, fome glafs-blowers, &c. and has
even been ufed a little by the chemifts and mineralifts; but, to the beft of my knowledge, Mr.
Cronftedt is the firft who made fuch an improvement in its ufe, as to be employed in examining
all mineral bodies. This gentleman invented fome
other apparatus, neceffary in making the experiments, to go with the Blow pipe, which all
together make a neat little cafe, that, for its
facility of being carried in the pocket, particularly
on travels, might be called a *Pocket-Laboratory:*
And as neither this Pocket Laboratory, nor even
the extenfive ufe of the Blow-pipe, is yet generally
known, I think it will not be altogether ufelefs, to
give a defcription of it.

SECT. IX.

The Blow-pipe is reprefented in its true figure
and fize, Tab. 1, fig. 1. The globe *a* is hollow,
and made on purpofe to condenfe the vapours,
which always happen to be in the Blow-pipe when
it has been ufed fome time: If this globe was not
there, the vapours would go directly with the
wind out into the flame, and would thereby cool
the affay.

The hole in the fmall end *b.* through which the
wind comes out, ought not to be larger than
the fize of the fineft wire. This hole may now
and then be ftopped up with fomething coming

into

into it, which hinders the force of the wind; one
ought therefore to have a piece of the fineſt wire,
to clear it with when required : And, in order to
have this wire the better at hand, it may be faſtened round the Blow-pipe, in ſuch a manner as is
repreſented in fig. 1. *c* is the wire, faſtened round
the Blow-pipe at *d.* and afterwards drawn through
a ſmall hole at *e.* made in the ring *f.* to keep it
more ſteady.

SECT. X.

The Blow-pipe is compounded of two parts,
Tab. 1. fig. 2. and 3 ; and this for the facility both
of making, carrying it along, and cleaning it on
the inſide when it is wanted.

In order to determine the moſt convenient proportions of this inſtrument, ſeveral Blow-pipes of
different ſizes, both bigger and ſmaller, have been
tried : The former have required too much wind,
and the latter being too ſoon filled with the wind,
have returned it back again upon the lungs :
Both theſe circumſtances hindered greatly the experiments, and are perhaps even prejudicial to the
health.   This ſize, fig. 1, is found to anſwer beſt ;
and though the hole muſt be as ſmall as before mentioned (Sect. ix.) yet the ſides of the pipe at the
point muſt not be thinner, nor the point narrower
than here repreſented, elſe it will be too weak,
and will not give ſo good a flame.   It is alſo to be
obſerved, that the canal throughout the pipe, but
particularly the hole at the ſmall end, muſt be
made very ſmooth, ſo that there are no inequalities
in it ; the wind would elſe be divided, and conſequently the flame made double.   That Blow-pipe
is to be reckoned the beſt, through which can be
formed the longeſt and moſt pointed flame from off a

common-

Tab. II.

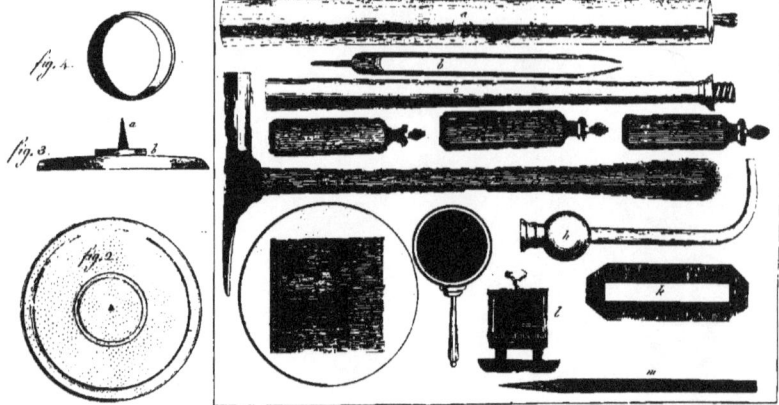

fig. 1.

fig. 4.

fig. 3.

fig. 2.

fig. 4.

fig. 3.

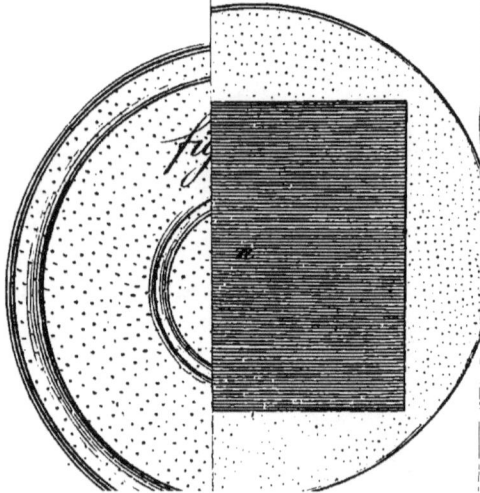

common-fized candle. Thefe Blow-pipes are commonly made of brafs or filver.

SECT. XI.

The whole Pocket-Laboratory is reprefented
Tab. II. fig. 1, with the cafe, exactly of the form,
bignefs, and proportions as that I make ufe of
myfelf: What alterations there may be wanted
are eafily found out by practice.

c h are the two parts of which the Blow-pipe
confifts, and which are already defcribed.

a. a wax-candle, deftined to be made ufe of,
particularly in travels, when no other candle is to
be had.

b. a pair of nippers, *(Korntong)* to handle fo
much the eafier the things which are to be tried,
becaufe they are generally fmall particles: This
ferves alfo to touch and turn the fubjects during
the experiments, when they are hot, and could
not be well handled with the fingers.

d. e. f. are three phials, to put the required
fluxes in, viz. Borax, the Mineral Alcali *(Sal
Sodæ)*, and *fal fufibile microcofmicum.*

g. a hammer, to break any part of a ftone,
when it is to be tried: This ferves alfo to pound
things with.

i. A magnifying glafs, neceffary when the objects are too fmall to be feen by the naked eye.

k. a fteel, to ftrike fire, by which the hardnefs
or foftnefs of the bodies is tried.

l. a loadftone, to difcover the prefence of iron.

m. a file, wherewith to diftinguifh natural gems,
quartz-criftals, and artificial or coloured glaffes
from one another.

n. a thin fquare plate made of untempered fteel,
filed flat on one fide, to pound things upon, and
polifhed

polifhed on the other fide, to hammer metals upon.

Above this fteel plate *n*. and within the circle, drawn round about it, is the place for a candleftick. This candleftick is fhewn in plan, fig. 2, and in profile, fig. 3, Tab. II. It confifts of a round brafs plate; the point *a*. and the ring *b*. round it, is inftead of the focket in another candleftick, which would here take up too much room.

Fig. 4, (Tab. II.) is a thin iron ring, a fixth part of an inch high; within this ring the pounding and grinding of the things upon the fteel plate fig. 1, *n*. is performed, that they may not be loft. In packing up, this ring is to be put loofe upon the candleftick; and, as it is lower than the point of this, it does not take up much room in the cafe.

The whole cafe, thus made, with all the inftruments in it, as I have defcribed them, is no more than one and an eighth part of an inch high, and confequently not more troublefome to be carried in the pocket than a fmall book *.

* The *Pocket Laboratory* here defcribed, and the *Box for the Acids*, mentioned in Sect. lxii. have been improved after the manner of Mr. Cronftedt, by a gentleman particularly acquainted with Mr. Engeftrom, from whom he learned this method of making Mineralogical Experiments. The bulk of the firft has been reduced nine and a half cubic inches; its length being diminifhed *one* fixteenth of an inch. the breadth *five* ditto, and the depth *two*; notwithftanding which, there is alfo added a piece of charcoal for trying the experiments, a flint, a piece of agaric tinder, and fome matches for lighting the candle. The three phials *d e f* for the falts, are of different colours, to prevent any miftake. The candleftick *s* has different concentric grooves for keeping the refults of the trials feparate. The blow-pipe *c h* has a filver mouth-piece, and fcrews in the middle of the ball, in order to clean out the moifture with the greater eafe; and the fmall wire [Sect. ix.] is more conveniently detached than fixed round it. The other box for

SECT. XII.

Whenever any thing is to be tried, one muſt not begin immediately with the Blow-pipe; ſome preliminary experiments ought to go before, by which thoſe in the fire may afterwards be directed. For inſtance, a ſtone is not always homogeneous, or of the ſame kind throughout, although it may appear to the eye to be ſo: The magnifying-glaſs is therefore neceſſary, to diſcover the heterogeneous particles, if there be any; and theſe ought to be ſeparated, and every thing tried by itſelf, that the effects of two different things, tried together, may not be attributed to one alone. This might happen with ſome of the finer *micæ*, which are now and then found mixed with ſmall particles of quartz, ſcarcely to be perceived by the eye. The Trapp, (in German *Schwartzſtein*) is alſo ſometimes mixed with very fine particles of Feltſpat (*ſpatum ſcintillans*) or of Calcareous Spar, &c. After this experiment follows that, to try the hardneſs of the ſtone in queſtion with the ſteel. The Flint and Garnet-kinds are commonly known to ſtrike fire with the ſteel; but there are alſo other ſtones, though very ſeldom, found ſo hard as to ſtrike fire:

for the acids, mentioned Sect. lxii. is reduced to leſs than a fourth of its original bulk, being exactly of the ſame ſize with the above. It contains two ſmall matraſſes [Tab. I. fig. 4.] for making ſolutions; a trough [Tab. I. fig. 5.] for waſhing the ore after its being pounded; and the three ſmall bottles with double ſtoppers, for the *nitrous, muriatic,* and *vitriolis* Acids, have their reſpective initials cut on each.

Both theſe Pocket-Laboratories, made in the neateſt manner by an ingenious artiſt, may be had ready furniſhed with the pureſt acids, &c. at the *General Office of Buſineſs, Arts and Trade,* opened for the preſent at No. 98, Wood-Street, Cheapſide, and only there, for very reaſonable prices.

A

184

A kind of Trapp is found of that hardnefs, in which no particles of Feltfpat are to be feen. Coloured glaffes refemble true gems ; but as they are very foft in proportion to thefe, they are eafily difcovered by the means of the file : The common quartz-cryftals are harder than coloured glaffes, but fofter than the gems. The loadftone difcovers the prefence of iron, when it is not mixed in too fmall a quantity in the ftone, and often before the ftone is roafted. Some kinds of *Hæmatites*, and particularly the *Cærulefcens*, is very like fome other iron ores, but diftinguifhes itfelf from thefe by a red colour, when pounded, the others giving a blackifh powder, and fo forth.

SECT. XIII.

To manage the Blow-pipe with eafe requires fome practice. A beginner blows generally too ftrongly, which forces him to take breath very often, and then he draws the flame at the fame time along into the Blow-pipe : This is trouble-fome for himfelf, and the experiment cools always a little at the fame time. But the more expe-rienced can breathe in, through the nofe, and yet at the fame time blow through the pipe, whereby a conftant flame from the candle is kept up. The whole art confifts in conftantly taking in air through the nofe, and with the tongue moderat-ing its blowing out; fo that the tongue performs nearly the office of a fucker in a pump; or rather, the action of the nofe, lungs, and mouth, refem-bles here the action of bellows with double parti-tions. In this manner there is no need of blowing violently, but only with a moderate and equal force, and thus the breath can never fail the ope-rator. The only inconveniency attending, is, that
the

the lips grow weak or tired, after having continued to blow for a while in one ftrain; but they foon recover their former ftrength, by ceafing to blow for fome minutes.

SECT. XIV.

The candle ufed for this purpofe (Seƈt. vii.) ought to be fnuffed often, but fo, that the top of the wick may retain fome fat in it, becaufe the flame is not hot enough when the wick is almoft burnt to afhes; but only the top muft be fnuffed off, becaufe a low wick gives too fmall a flame. The blue flame is the hotteft; this ought therefore to be forced out when a great heat is required, and only the point of the flame muft be direƈted upon the fubjeƈt which is to be affayed.

SECT. XV.

The piece of charcoal made ufe of in thefe experiments (Seƈt. vii.), muft not be of a difpofition to crack. If this fhould happen, it muft gradually be heated until it does not crack any more, before any affay is made upon it. If this is not obferved, but the affay made immediately with a ftrong flame, fmall pieces of it will fplit off in the face and eyes of the affayer, and often throw along with them the matter that was to be affayed. Charcoal which is too much burnt confumes too quick during the experiment, leaving fmall holes in it, wherein the matter to be tried may be loft: And charcoal that is burnt too little catches flame from the candle, burning by itfelf like a piece of wood, which likewife hinders the procefs.

SECT.

SECT. XVI.

Of thofe things that are to be affayed, only a fmall piece muft be broke off for that purpofe, not bigger than that the flame of the candle (Sect. vii. xiv.) may be able to act upon it at once, if required; which is fometimes neceffary; for inftance, when the matter requires to be made red hot throughout. A piece of about an eighth part of an inch fquare is reckoned of a moderate fize, and fitteft for experiments; feldom more, but rather lefs. This proportion is only mentioned as a direction in regard to the quantity, the figure being of no confequence at all, a piece broke off from a ftone feldom or never happening to be fquare. But here it is to be obferved, that the piece ought to be broke as thin as poffible, at leaft the edges: The advantage thereof is eafily feen, the fire having then more influence upon the fubject, and the experiment being quicker made. This is particularly neceffary to be obferved when fuch ftones are to be affayed, which although in fome refpects fufible by themfelves, yet refift confiderably the action of the fire; becaufe they may by thefe means be brought into fufion, at leaft at their edges, which elfe would have been very difficult if the piece had been thick.

SECT. XVII.

Some of the mineral bodies are very difficult to keep fteady upon the charcoal during the experiment, before they are made red hot; becaufe, as foon as the flame begins to act upon them, they fplit afunder with violence, and difperfe. Such often

often are thofe which are of a foft confiftence, or a particular figure, and which preferve the fame figure in however minute particles they are broke; for inftance, the Calcareou. Spar, the Sparry Gypfum, Sparry Fluor, White Sparry Lead-ore, the Potters Ore, (*Galena teffellata*) the Teffellated Mock-lead or Blende, &c. even all the common fluors which have no determinate figure, and moft of the *Minera metallorum calciformes cryftallifata* or *fpatofa*: All thefe are not fo compact as common hard ftones; and therefore, when the flame is immediately pufhed at them, the heat forces itfelf quickly through and into their clefts or pores, and caufes this violent expanfion and difperfion. Many of the clays are likewife apt to crack in the fire, which may be for the moft part afcribed to the humidity, of which they always retain a portion. Befides thefe enumerated, there may be found now and then other mineral bodies of the fame nature; but it is, however, not fo common.

The only way of preventing this inconveniency, is to heat the body as flowly as poffible. It is beft, firft of all, to heat that place of the charcoal, where the piece is intended to be put on, and afterwards lay it thereon; a little crackling will then enfue, but commonly of no great confequence. After that, the flame is to be blown very flowly towards it, in the beginning not directly upon, but fomewhat above it, and fo approaching nearer and nearer with the flame until it becomes red hot. This will do for the moft part; but there are neverthelefs fome, which, notwithftanding all thefe precautions, it is almoft impoffible to keep on the charcoal. Thus the Fluors are generally the moft difficult; and as one of their principal characters is difcovered by their effects in the fire *per fe*, (Sect. xviii. 6.) they ought neceffarily

ceffarily to be tried that way. To this purpofe it is beft to make a little hole in the charcoal to put the Fluor in, and then to put another piece of charcoal as a covering upon this, leaving only a fmall opening for the flame to come in at, and to look at the proof. As this ftone will neverthelefs moftly fplit and fly about, a larger piece thereof than is before-mentioned, (Sect. xvi.) muft be taken, in order to have at leaft fomething of it left.

But if the experiment is to be made upon a ftone whofe effects one does not want to fee in the fire *per fe,* but rather with fluxes, then a piece of it ought to be forced down into melted borax, (Sect. xxiii.) when always fome part of it will remain in the borax, notwithftanding the greateft part may fometimes fly away by cracking.

S E C T. XVIII.

As the ftones undergo great alterations when expofed to the fire by themfelves, whereby fome of their characteriftics, and often the moft principal, are difcovered, they ought firft to be tried that way; obferving what has been faid before concerning the quantity of the matter, direction of the fire, &c. The following effects are generally the refults of this experiment, viz.

1. Calcareous earth or ftone, when it is pure, does never melt by itfelf, but becomes white and friable, fo as to break freely between the fingers; and, if fuffered to cool, and then mixed with water, it becomes hot, juft as common quick lime. As in thefe experiments only very fmall pieces are ufed, (Sect. xvi.) this laft effect is beft difcovered by putting the proof on the outfide of the hand, with a drop of water to it, when inftantly a very

quick

quick heat is felt on the ſkin. When the calcareous ſubſtance is mixed with the vitriolic acid,
as in the gypſum; or with a clay, as in the marle;
it commonly melts by itſelf; yet more or leſs difficult in proportion to the differences of the mixtures: The gypſum produces generally a white,
and the marle a grey glaſs or flag. When there is
any iron in it, as in white iron ore, it becomes
dark, and ſometimes quite black, &c.

2. The Siliceæ never melt alone, but become
generally more brittle after being burnt: Such of
them as are coloured become colourleſs, and the
ſooner when it does not ariſe from any contained
metal; for inſtance, the Topazes, Amethiſts, &c.
ſome of the precious ſtones, however, excepted.
And ſuch as are mixed with a quantity of iron,
grow dark in the fire, as ſome of the Jaſpers, &c.

3. The Garnet-kind melt always into a black
flag, and that ſometimes ſo eaſy, that it may be
brought into a round globule upon the charcoal.

4. The Argillaceæ, when pure, never melt,
but become white and hard: The ſame effects
follow when they are mixed with phlogiſton; for
inſtance, the *Soap-rock* is eaſily cut with the
knife; but, being burnt, it cuts glaſs, and would
ſtrike fire with the ſteel, if as large a piece, as is
neceſſary for that purpoſe, could be tried in this
way. The *Soap-rocks* are ſometimes found of a
dark brown and nearly black colour, but become
for all that quite white in the fire, as a piece of
China ware: However, care muſt be taken not to
puſh the flame from the top of the wick, there
being for the moſt part a ſooty ſmoke, which commonly will darken all that it touches; and if this
is not obſerved, a miſtake in the experiment might
eaſily happen: But if it is mixed with iron, as it
is ſometimes found, it does not ſo eaſily part with

U   its

its dark colour. The Argillaceæ, when mixed with lime, melt by themfelves, as above-mentioned (1). When mixed with iron, as in the Boles, they grow dark or black; and if the iron is not in too great a quantity, they melt alone into a dark flag; the fame happens, when they are mixed with iron and a little of the vitriolic acid, as in the common clay, &c.

5. The Micaceæ and Afbeftinæ become fomewhat hard and brittle in the fire, and are more or lefs refractory, though they give fome marks of fufibility.

6. The Fluores difcover one of their chief characteriftics by giving a light, like Phofphorus, in the dark, when they are flowly heated; but lofe this property, as well as their colour, as foon as they are made red hot: They commonly melt in the fire into a white opaque flag, though fome of them not very eafily.

7. Some forts of the Zeolites, a ftone lately difcovered, melt eafily and foam in the fire, fometimes nearly as much as Borax, and become a frothy flag, &c.

8. A great many of thofe mineral bodies which are impregnated with iron, as the Boles, and fome of the White Iron Ores, &c. as well as fome of the other iron ores, viz. the Bloodftone, are not attracted by the loadftone before they have been thoroughly roafted, &c.

A further digreffion upon thefe effects is unneceffary here, their enumeration belonging more properly to the Mineralogy; it is fufficient only to have mentioned the moft common, in order the better to explain the experiments that are made with the Blow-pipe.

SECT.

SECT. XIX.

After the mineral bodies have been tried in the fire by themfelves, they ought to be melted with fluxes, to find out if they can be diffolved or not, and fome other phœnomena attending this operation. To this purpofe three different kinds of falts are ufed as fluxes, viz. *Sal Sodæ, Borax,* and *Sal fufible microcofmicum* (Sect. xi).

SECT. XX.

The Sal Sodæ is a mineral alcali well known, prepared from the herb Kali or Saltwort; this falt is however not much ufed in thefe fmall experiments, its effects upon the charcoal rendering it, for the moft part, unfit for it; becaufe, as foon as the flame begins to act upon it, it melts inftantly, and is almoft wholly attracted by the charcoal. When this falt is employed to make any experiment, but a very little quantity thereof is wanted at once, viz. about the cubical contents of an eighth part of an inch, more or lefs : This is laid upon the charcoal, and the flame blown on it with the Blow-pipe ; but as this falt commonly is in form of a powder, it is neceffary to go on very foftly, that the force of the flame may not difperfe the minute particles of the falt. As foon as it begins to melt it runs along on the charcoal almoft as melted tallow, and when cold, it is a glaffy matter of an opaque dull colour fpread on the coal. The moment it is melted the matter which is to be tried ought to be put into it, becaufe otherwife the greateft part of the falt will be foaked into the charcoal, and too little of it left for the intended purpofe ; the flame ought

then

then to be directed on the matter itfelf, and if
the falt fpreads too much about, leaving the proof
almoft alone, it may be brought to it again by
blowing the flame on its extremities, and direct-
ing it towards the fubject of the experiment.
In the affays made with this falt, it is true, we
may find if the mineral bodies which are melted
with it have been diffolved by it or not ; but we
cannot tell with any certitude whether this is done
haftily and with force, or gently and flowly ;.
whether only a lefs or a greater part of the matter
has been diffolved ;. nor can it be well diftinguifhed
if the matter has imparted any weak tincture to
the flag ; becaufe this falt always bubbles upon
the charcoal during the experiment, nor is it clear.
when cool ; fo that fcarce any colour, except it be
a very deep one, can be difcovered, although
it may fometimes be coloured by the matter that
has been tried.

S E C T. XXI.

The other two falts, viz. the Borax, and the
Sal fufible microcofmicum, are very well adapted
to thefe experiments, becaufe they may by the
flame be brought to a clear uncoloured and tranfparent glafs ; and as they have no attraction to
the charcoal, they keep themfelves always upon
it in a round globular form. The Sal fufibile microcofmicum is very fcarce, and perhaps not to
be met with in the fhops ; it is made of urine :
Mr. Margraff has given a full account of its preparation in the Memoirs of the Academy of
Sciences at Berlin.

S E C T.

S E C T. XXII.

The quantity of thefe two falts required for an experiment is almoft the fame as the Sal Sodæ (Sect. xx.); but as thefe falts are criftallifed, and confequently include a great deal of water, particularly the borax, their bulk is confiderably reduced when melted, and therefore a little more of thefe may be taken than the before-mentioned quantity.

S E C T. XXIII.

Both thefe falts, (Sect. xxi.) when expofed to the flame of the Blow pipe, bubble very much and foam before they melt to a clear glafs, but more fo the borax, which for the moft part depends on the water they contain: And as this would hinder the affayer to make due obfervations on the phœnomena of the experiment, the falt which is to be ufed, muft firft be brought to a clear glafs, (Sect. xxi.) before it can ferve as a flux; it muft therefore be kept in the fire until it is become fo tranfparent that the cracks in the charcoal may be feen through it. This done, whatfoever is to be tried, is put to it, and the fire continued.

S E C T. XXIV.

Here it is to be obferved, that for the affays made with any of thefe two fluxes (Sect. xxii) on mineral bodies, no larger pieces of thefe muft be taken, than that altogether they may keep a globular form upon the charcoal; becaufe then it may be better diftinguifhed in what manner the flux acts upon the matter during the ex-

U 3　　　　　- periment:

periment : If this is not obferved, the flux, communicating itfelf with every point of the furface of the mineral body, fpreads all over it, and keeps the form of this laft, which commonly is flat, (Sect. xvi.) and by that means hinders the operator to obferve all the phœnomena which may happen. Befides, the flux being in too fmall a quantity, in proportion to the body to be tried, is too weak to act with all its force upon it. The beft proportion, therefore, is about a third part of the mineral body to the flux; and, as the quantity of the flux, mentioned in Sect. xx. xxii. makes a globe of a due fize, in regard to the greateft heat that is poffible to procure in thefe experiments; the fize of the mineral body, propofed in Sect. xvi. required when it is to be tried in the fire by itfelf, is too large on this occafion, the third part of it being here almoft fufficient.

SECT. XXV.

The Sal Sodæ, as has been faid before, is not of much ufe in thefe experiments; nor has it any particular qualities in preference to the two laft mentioned falts, except that it diffolves the Zeolites eafier than the Borax and the Sal fufibile microcofmicum.

This laft mentioned falt fhews almoft the fame effects in the fire as the borax, and differs from this in very few circumftances, of which one of the moft principal is, that, when melted with manganefe, it becomes of a crimfon hue, inftead of a jacinth colour, which borax takes.

This falt is, however, for its fcarcity, ftill very little in ufe, borax alone being that which is commonly ufed. Whenever a mineral body is melted with any of thefe two laft mentioned falts, in the above

above defcribed manner (Sect. xxii. *et feq.*) it is easily feen whether it is quickly diffolved, becaufe in that cafe an effervefcence arifes, which lafts till the whole is diffolved; or whether this is flowly done, in which cafe few and fmall bubbles only rife from the matter : Likewife, if it cannot be diffolved at all, becaufe then it is obferved only to turn round in the flux without the leaft bubble, and the edges look as fharp as they were before.

S E C T. XXVI.

In order further to illuftrate what has been faid about thefe experiments, I will mention fome inftances out of the Mineralogy, concerning the effects of borax upon the mineral bodies, viz.

1. The calcareous fubftances, and all thofe ftones which contain any thing of lime in their compofition, diffolve readily and with effervefcence in the borax : This effervefcence is the more violent, the greater the portion of lime contained in the ftone. This reafon, however, is not the only one in the gypfum, becaufe both the conftituents of this do readily mix with the borax, and therefore a greater effervefcence arifes in melting gypfum with the borax, than lime alone.

2. The Siliceæ do not diffolve, unlefs fome few, which contain a quantity of iron.

3. The Argillaceæ, when pure, are not acted upon by the borax; but when they are mixed with fome heterogeneous bodies, they are diffolved, though very flowly; fuch is for inftance the Stone Marrow, the Common Clay, &c.

4. The Granateæ, Zeolites, and Trapp, diffolve but flowly.

5. The Fluores, Afbeftinæ, and Micaceæ, diffolve for the moft part very eafily, and fo forth.

S E C T.

S E C T. XXVII.

Some of thefe bodies melt to a colourlefs tranf-parent glafs with the borax; for inftance, the Calcareous Subftances, when pure, the Fluores, fome of the Zeolites, &c. Others tinge the borax with a green tranfparent colour; viz. the *Granatea*, *Trapp*, fome of the *Argillacea*, fome of the *Micacea* and *Afbeftina*: This green has its origin, partly from a fmall portion of iron, which the *Granatea* particularly contain, and partly from phlogifton.

S E C T. XXVIII.

The borax cannot diffolve but a certain quantity of a mineral body proportional to its own: Of the calcareous kind it diffolves a vaft quantity, but turns at laft, when too much has been added, from a clear, tranfparent, to a white, opaque flag. When the quantity of the calcareous matter exceeds but little in proportion, the glafs looks very clear as long as it remains hot; but as foon as it begins to cool, a white half opaque cloud is feen to arife from the bottom, which fpreads over the third, half, or more of the glafs globe, in proportion to the quantity of calcareous matter; but the glafs or flag is neverthelefs fhining, and of a glaffy texture when broke; if more of this matter be added, the cloud rifes quicker and more opaque, and fo by degrees till the flag becomes quite milk white: It is then no more of a fhining, but rather dry appearance, on the furface; is very brittle, and of a grained texture, when broke.

SECT. XXIX.

All that has been faid hitherto of experiments upon mineral bodies, is only concerning the ftones and earths.   I am now proceeding to the metals and ores, in order to defcribe the manner of examining thefe bodies, and particularly the management of the Blow-pipe in thefe experiments, An exact knowledge, and nice proceeding are fo much the more neceffary here, as the metals are often fo difguifed in their ores, as to be very difficultly known by their external appearance, and liable fometimes to be miftaken one for the other: Some of the cobalt ores for inftance, refemble much a Pyrites Arfenicalis; there are alfo fome iron and lead ores, which are nearly like one another, &c.

SECT. XXX.

As the ores generally confift of metals mineralifed with fulphur or arfenic, or fometimes both together; they ought firft to be expofed to the fire by themfelves, in order, not only to determine with which of thefe they are mineralifed, but alfo to fet them free from thefe volatile mineralifing bodies: Thus this ferves inftead of calcination, by which they are prepared for further effays.

SECT. XXXI.

Here it muft be obferved, that, whenever any metal, or fufible ore is to be tried, a little concavity muft be made in that place of the charcoal where the matter is to be put; becaufe, as foon as it is melted, it forms itfelf into a globular figure, and

and might then roll from the charcoal, if its surface was plain; but when borax is put to it, this inconveniency is not so much to be feared.

SECT. XXXII.

Whenever an ore is to be tried, a small bit is broke off for that purpose, of such a size as is directed in Sect. xvi. this bit is laid upon the charcoal, and the flame blown on it slowly: Then the sulphur or arsenic begins to part from it in form of smoke; these are easily distinguished from one another by their smell, that of sulphur being sufficiently known, and the arsenic smelling like garlick. The flame ought to be blown very softly, as long as any smoke is seen to part from the ore; but, after that, the heat must be augmented by degrees, in order to make the calcination as perfect as possible. If the heat is applied very strong from the beginning upon an ore, that contains much of the sulphur, or arsenic, this ore will presently melt, and yet lose very little of its mineralising bodies, and by that means render the calcination very imperfect. It is however, impossible to calcine the ores in this manner to the utmost perfection, which is easily seen in the following instance, viz. in melting down a calcined Potter's ore with borax, it will be found to bubble upon the coal, which depends on the sulphur, which is still left, the vitriolic acid of this uniting with the borax, and causing this motion. However, lead in its metallic form, melted in this manner, bubbles alone upon the charcoal, if any sulphur remains in it. But, as the lead, as well as some of the other metals, may raise bubbles upon the charcoal, although they are quite free from the sulphur, only by the flames being forced too violently

lently on it, thefe phœnomena ought not to be confounded with each other.

SECT. XXXIII.

The ores being thus calcined, the metals contained in them may be difcovered, either by being melted alone, or with fluxes : when they fhew themfelves, either in their pure metallic ftate, or by tinging the flag with colours peculiar to each of them. In thefe experiments it is not to be expected, that the quantity of metal contained in the ore fhould be exactly determined; this muft be done in larger laboratories. This cannot, however, be looked upon as any defect, fince it is fufficient for a mineralift, only to find out what fort of metal is contained in the ore. There is another circumftance, which I am forry to fay, is a more real defect in our little laboratory, which is, that fome ores are not at all able to be tried in it, by fo fmall an apparatus : for inftance, the gold ore called *Pyrites aureus*, which confifts of gold, iron, and fulphur. The greateft quantity of gold, which this ore contains, is about one ounce, or one ounce and an half out of one hundred pounds of the ore, the reft being iron and fulphur ; and as only a very fmall bit is allowed for thefe experiments, (Sect. xvi. xxxi.) the gold contained therein, can hardly be difcerned by the eye, even if it could be extracted, but it goes along with the iron in the flag, this laft metal being in fo large a quantity in proportion to the other, and both of them having a commicible power with each other.

All the kinds of Blende, Black jack, which are mineralifed zink ores, containing zink, fulphur, and iron, cannot be tried this way, becaufe they cannot be perfectly calcined, and befides, the zink

flies

flies off, when the iron fcorifies: neither can all thofe Blendes, which contain filver or gold mineralifed with them, be tried in this manner, which is particularly owing to the imperfect calcination ; nor are the quickfilver ores fit for thefe experiments, the volatility of this femi-metal making it impoffible to bring it out of the poorer fort of ores * ; and the rich ores, which fweat out the quickfilver, when kept clofe in the hand, not wanting any of thefe affays, &c. Thofe ores ought to be affayed in larger quantities, and even with fuch other methods, as cannot be applied upon a piece of charcoal.

S E C T. XXXIV.

Some of the rich filver ores are eafily tried : for inftance, *Minera argenti vitrea*, commonly called Silver-glafs, which confifts only of filver and fulphur. When this ore is expofed to the flame, it melts inftantly, and the fulphur goes away in fume, leaving the filver pure upon the charcoal, in a globular form. If this filver fhould happen to be of a dirty appearance, which often is the cafe, then it muft be melted anew with a very little borax, and after it has been kept in fuffion for a minute or two, fo as to be perfectly melted and red-hot, the proof is fuffered to cool : it may then be taken off the coal, and being laid upon the fteel-plate, (Sect. xi. *n.*) the filver is feparated from the flag by one or two ftrokes of the hammer (Sect. xi. *g.*). Here the ufe of the iron ring (Sect. xi.) is manifeft, for this ought firft to be placed upon the plate, to hinder the proof from flying off by the violence of the ftroke, which otherwife would

* A piece of gold being laid over the proof, to receive the fumes, readily difcovers if it contains any quick-filver. And it is probable, that by like proceffes, we may alfo be enabled to difcover with the Blow-pipe other of the volatile fubftances.

happen.

happen. The filver is then found inclofed in the
flag of a globular form, and quite fhining, as if it
were polifhed. When a large quantity of filver is
contained in a lead ore, viz. in a potter's ore, it
can likewife be difcovered through the ufe of the
blowpipe, of which more will be mentioned hereafter. (Sect. xxxix.)

S E C T. XXXV.

Of the pure Tin ores, the tin may be melted
out in its metallic ftate. Some of thefe ores melt
very eafily, and yield their metal in quantity, if
only expofed to the fire by themfelves: but others
are more refractory, and as thefe melt very flowly,
the tin, which fweats out in form of very fmall glo-
bules, is inftantly burnt to afhes, before thefe glo-
bules have time to unite, in order to compofe a
larger globe, which might be feen by the eye, and
is not fo foon deftroyed by the fire; it is therefore
neceffary to add a little borax to thefe from the
beginning, and then to blow the flame violently
at the proof. The borax does here preferve the
metal from being too foon calcined, and even contributes to the readier collecting of the fmall metallic particles, which foon are feen to form themfelves into a globule of metallic tin at the bottom
of the whole mafs, neareft to the charcoal. As
foon as fo much of metallic tin is produced, as is
fufficient to convince the operator of its prefence,
the fire ought to be ceafed, although not the whole
of the ore is yet melted; becaufe feldom, or rather never, the whole of this kind of ore can be
reduced into metal by means of thefe experiments,
a great deal thereof always being calcined: and if
the fire is continued too long, perhaps even the
metal, already reduced, may likewife be burnt to
afhes: for the tin is very foon deftroyed from its
metallic ftate by the fire. S E C T.

S E C T. XXXVI.

Moſt part of the lead ores may be brought to a metallic lead upon the charcoal. The *Mineræ plumbi calciformes*, which are pure, are eaſily melted into lead : but ſuch of them, as are mixed with an *ochraferri*, or any kind of earth, as Clay, Lime, &c. yield very little of lead, and even nothing at all, if the heterogenea are combined therewith in any large quantity : this happens even with the *Minera plumbi calciformis arſenico mixta*. Theſe, therefore, are not to be tried but in larger laboratories. However, every mineral body ſuſpected to contain any metallic ſubſtance, may be tried by the blow-pipe, ſo as to give ſufficient proofs, whether it contains or not, by its effects being different from thoſe of the ſtone or earths, &c.

S E C T. XXXVII.

The *Mineræ plumbi mineraliſatæ,* leave the lead in a metallic form, if not too large a quantity of iron is mixed with it. For example, when a teſſellated or ſteel-grained lead ore is expoſed to the flame, its ſulphur, and even the arſenic, if there be any, begins to fume, and the ore itſelf immediately to melt into a globular form ; the reſt of the ſulphur continues then to fly off, if the flame is blown ſlowly upon the maſs, inſtead of that, very little of the ſulphur will go off, if the flame is forced violently on it : in this caſe, it rather happens that the lead itſelf crackles and diſſipates, throwing about very minute metallic particles. The ſulphur being driven out, as much as poſſible, which is known by finding no ſulphureous vapour in ſmelling at the proof, the whole is ſuffered

fered to cool, and then a globule of metallic lead
will be left upon the coal. If any iron is contained
in the lead ore, the lead, which is melted out of
it, is not of a metallic fhining, but rather of a
black and uneven furface : a little borax muft in
this cafe be melted with it, and as foon as no bubble is feen to rife any longer from the metal into
the borax, the fire muft be difcontinued : when
the mafs is grown cold, the iron will be found
fcorified with the borax, and the lead left pure,
and of a fhining colour.

SECT. XXXVIII.

The borax does not fcorify the lead in thefe
fmall experiments, when it is pure : if the flame
is forced with violence on it, a bubbling will enfue,
refembling that which is obferved when borax
diffolves a body melted with it, but when the fire
ceafes, the flag will be perfectly clear and tranfparent, and a quantity of very minute lead particles
will be feen fpread about in the borax, which have
been torn off from the mafs during the bubbling.

SECT. XXXIX.

If fuch a lead ore (Sect. xxxvii.) is rich in filver,
this laft metal may likewife be difcovered by this
experiment; becaufe, as the lead is volatile, it
may be forced off, and the filver remain. To effect
this, the lead, which is melted out of the ore,
muft be kept in conftant fufion with a flow heat,
that it may be confumed. This end will be fooner
obtained, and the lead part quicker, if, during
the fufion, the wind through the Blow-pipe is directed immediately, though not forcibly, upon the
melted mefs itfelf, until it begins to cool, then the
fire muft be directed on it again. The lead, which

is

is already in a volatilifing ftate, will by this artifice
be driven out in form of a fubtil fmoke; and by thus
continuing by turns, to melt the mafs, and then to
blow off the lead, as has been faid, until no fmoke
is any longer perceived, the filver will at laft be
obtained pure. The fame obfervation holds good
here alfo, which was made about the gold, that,
as none but very little bits of the ores can be employed in thefe experiments, it will be difficult to
extract the filver out of a poor ore; for fome part
of it will fly off with the lead, and, what might
be left, is too little to be difcerned by the eye.
The filver, which, by this means is obtained, is
eafily diftinguifhed from lead by the following external marks, viz. that it muft be red-hot, before
it can be melted: it cooles fooner than lead: it
has a filver colour; that is to fay, brighter and
whiter than lead: and is harder to beat with the
hammer. (Sect. xxxiv.)

SECT. XL.

The *Mineræ cupri calciformes*, (at leaft fome of
them) when not mixed with too much ftone or
earth, are eafily reduced to copper with any flux:
if the copper is found not to have its natural
bright colour, it muft be melted with a little borax,
which purifies it. Some of thefe ores do not at all
difcover their metal, if not immediately melted
with borax; the heterogenea, contained in them,
hindering the fufion, before thefe are fcorified by
the flux.

SECT. XLI.

The grey Copper ores, which only confift of
copper and fulphur, are tried almoft in the fame
manner, as above-mentioned. (Sect. xl.) Being
expofed

expofed to the flame by themfelves, they will be found inftantly to melt, and part of their fulphur to go off; the copper may afterwards be obtained in two ways: the one, by keeping the proof in fufion for about a minute, and after fuffering it to cool; when it will be found to have a dark and uneven appearance externally, but which, after being broke, difcovers the metallic copper of a globular form in its centre, furrounded with a regule, which ftill contains fome fulphur and a portion of the metal: the other, by being melted with borax, which laft way fometimes makes the metal appear fooner.

S E C T. XLII.

The *Minera cupri pyritacea*, containing copper, fulphur, and iron, may be tried with the blowpipe, if they are not too poor: in thefe experiments the ore ought to be calcined, and, after that, the iron fcorified. For this purpofe a bit of the ore muft be expofed to a flow flame, that as much of the fulphur as poffible may part from it, before it is melted, becaufe the ore commonly melts very foon, and then the fulphur is more difficult to drive off. After being melted, it muft be kept in fufion with a ftrong fire, for about a minute, that a great part of the iron may be calcined: and, after that, fome borax muft be added, which fcorifies the iron, and turns with it to a black flag. If the ore is very rich, a metallic copper will be had in the flag, after the fcorification: if the ore is of a moderate richnefs, the copper will ftill retain a little fulphur, and fometimes iron: the product will therefore be brittle, and muft with great caution be feparated from the flag, that it may not break into pieces; and if this product is

X   after-

afterwards treated in the fame manner as before faid, in fpeaking of the grey copper-ores (Sect. xli.), the metal will foon be produced. But, if the ore is poor, the product after the firft fcorification muft be brought into fufion, and afterwards melted with fome frefh borax, in order to calcine and fcorify the remaining portion of iron ; after which it may be treated as mentioned in Sect. xli. The copper will, in this laft cafe, be found in a very fmall globule.

S E C T. XLIII.

The copper is not very eafily fcorified with this apparatus, when it is melted together with borax ; unlefs it has firft been expofed to the fire by itfelf for a while, in order to be calcined. When only a little of this metal is diffolved, it inftantly tinges the flag of a reddifh brown colour, and moftly opaque ; but as foon as this flag is kept in fufion for a little while, it becomes quite green and tranfparent : and thus the prefence of the copper may be difcovered by the colour, when it is concealed in heterogeneous bodies, fo as not to be difcovered by any other experiment.

S E C T. XLIV.

If metallic copper is melted with borax by a flow fire, and only for a very little time, the glafs, or flag, becomes of a fine tranfparent blue or violet colour, inclining more or lefs to the green ; but this colour is not properly owing to the copper, but it may rather be to its phlogifton ; becaufe the fame colour is to be had in the fame manner from iron : and thefe glaffes, which are coloured with either of thefe two metals, foon lofe their colour, if expofed

poſed to a ſtrong fire, in which they are made quite
clear, and colourleſs. Beſides, if this glaſs, tinged
blue with the copper, is again melted with more
of this metal, it becomes of a good green colour,
which for a long time keeps unchanged in the fire.

SECT. XLV.

The iron ores, when pure, can never be melted
by themſelves, through the means of the blow-
pipe alone, nor do they yield their metal, when
melted with fluxes, becauſe they require too ſtrong a
heat to be brought into fuſion ; and, as both the ore
and the metal itſelf very ſoon loſe their phlogiſton
in the fire, and cannot be ſupplied with a ſufficient
quantity from the charcoal, ſo likewiſe they are
very ſoon calcined in the fire. This eaſy calcination is alſo the reaſon why the fluxes, for inſtance borax, readily ſcorify this ore, and
even the metal itſelf. The iron loſes its phlogiſton in the fire ſooner than the copper, it is therefore eaſier ſcorified ; and this is the principle on
which the experiment mentioned in Sect. xlii. is
founded.

SECT. XLVI.

The iron is, however, diſcovered without much
difficulty, although it were mixed but in a very
ſmall quantity with heterogeneous bodies. The
ore, or thoſe bodies which contain any large
quantity of the metal, are all attracted by the loadſtone, ſome without any previous calcination, and
others not till after having being roaſted. When
a clay is mixed with a little iron, it commonly
melts by itſelf in the fire ; but, if this metal is
contained in a limeſtone, it does not promote the

fuſion,

fufion, but gives the ftone a dark, and fometimes a deep black colour, which always is the character of iron. A *Minera ferri calciformis pura cryftallifata*, is commonly of a red colour: This being expofed to the flame, becomes quite black, and is then readily attracted by the loadftone, which it was not before. Befides thefe figns, the iron difcovers itfelf, by tinging the flag of a green tranfparent colour, inclining to brown, when only a little of the metal is fcorified; but as foon as any larger quantity thereof is diffolved in the flag, this becomes firft a blackifh brown, and afterwards quite black and opaque.

SECT. XLVII.

Bifmuth is known by its communicating a yellowifh brown colour to borax: and Arfenic by its volatility, and garlick fmell. Antimony, both in form of regulus and ore, is wholly volatile in the fire, when it is not mixed with any other metal (except arfenic), and is known by its particular fmell; eafier to be diftinguifhed, when once known, than defcribed. When the ore of antimony is melted upon the charcoal, it bubbles conftantly, during its volatilifing.

SECT. XLVII.

Zinc ores are not eafily tried upon the coal (Sect. xxxiii.). But the regulus of zinc, expofed to the fire upon the charcoal, burns with a beautiful blue flame, and forms itfelf almoft inftantly into white flowers, which are the common flowers of zinc.

SECT.

S E C T. XLVIII.

Cobalt is particularly remarkable for giving to the glafs a blue colour, which is the zaffre or fmalt. To produce this, a piece of cobalt ore muft be calcined in the fire (Sect. xxx. xxxi.) and afterwards melted with borax. As foon as the glafs, during the fufion, from being clear, feems to grow opaque, it is a fign, that it is already tinged a little ; the fire is then to be difcontinued, and the operator muft take hold with the nippers (Sect. xi. 6.) of a little of the glafs, whilft yet hot, and draw it out flowly in the beginning, but afterwards very quick, before it cools, whereby a thread of the coloured glafs is procured, more or lefs thick, on which the colour may eafier be feen againft the day or candle-light, than if it was left in a globular form. This thread melts eafily if only put in the flame of the candle, without the help of the blow-pipe.

If this glafs is melted again with more of the cobalt, and kept in fufion for a while, the colour becomes very deep ; and thus the colour may be altered, according to pleafure.

S E C T. L.

When the cobalt ore is pure, or at leaft contains but little iron, a cobalt regulus is almoft inftantly produced in the borax, during the fufion : but when it is mixed with a quantity of iron, this laft metal ought firft to be feparated, which is eafily performed, fince it fcorifies fooner than the cobalt ; therefore, as long as the flag retains any brown or black colour Sect. xlviii. it muft be feparated,

and

and melted again with fresh borax, until it shews the blue colour.

SECT. LI.

Nickel is very seldom to be had, and as its ores are seldom free from mixtures of other metals, it is very difficultly tried with the blowpipe. However, when this semi-metal is mixed with iron and cobalt, it is easily freed from these heterogeneous metals, and reduced to a pure nickel regulus by means of scorification with borax, in the same manner as is mentioned Sect. l. because both the iron and cobalt sooner scorify than the nickel. The regulus of nickel itself is of a green colour, when calcined: it requires a pretty strong fire before it melts, and tinges the borax with a jacint colour. Manganese gives the same colour to borax, but its other qualities are quite different, so as not to be confounded with the nickel.

SECT. LII.

Thus I have briefly described the use of the Blow-pipe, and the method of employing it in the study of Mineralogy. Any gentleman who is a lover of this science, will, by attending to the rules here laid down, be able in an easy manner to amuse himself in discovering the properties of those works of nature which the mineral kingdom furnishes us with. The husbandman may by its help find out what sorts of stones, earths, ores, &c. there are on his estate, and to what œconomical uses they may be employed. The Scientific Mineralist may, by examining into the properties and effects of the mineral bodies, discover the natural relation

relation thefe bodies ftand in to each other, and thereby furnifh himfelf with materials for eftablifh-ing a Mineral Syftem, founded on fuch principles as Nature herfelf has laid down in them; and this in his own ftudy, without being forced to have re-courfe to great laboratories, crucibles, furnaces, &c. which is attended with a great deal of trouble, and is the reafon why fo few can have an op-portunity of gratifying their defire of knowledge in this part of natural hiftory. I do not pretend to fay, that the Pocket-Laboratory here defcribed, is in every refpect as perfect as it is capable of being made : and I have in the foregoing pages indicated fome inftances where it is not fufficient; yet thofe inftances are very few. Befides, the fhort time fince it has been invented, and the few perfons that have known how to make ufe of it, are a fufficient apology for its not being brought to the utmoft degree of perfection. It is to be hoped, that the more general its ufe be made, the more and fooner will fuch imperfections be removed, and fuch wants filled up, as will be found neceffary and convenient. I fhall now add fome hints towards thefe improvements, leaving to the judicious prac-titioner the manner of completing them.

SECT. LIII.

A greater number of fluxes might, perhaps, be found out, whofe effects on mineral bodies might be different from thefe already in ufe, whereby more diftinct characters of thofe mineral bodies might be difcovered, which now either fhew am-biguous ones, or which are almoft impoffible to try exactly with the Blow-pipe. Inftead of the *fal foda*, fome other falts might be found out, more fit for thefe experiments. But it is very neceffary

X 4

not

not to make ufe of any other fluxes than fuch as have no attraction to the charcoal: if they, at the fame time, are clear and tranfparent, when melted, as the borax and the *fal fufibile microcof-micnm*, it is ftill better : however, the tranfparency or opacity are of no great confequence, if a fub-ftance is effayed only in order to difcover its fufi-bility, without any attention to its colour ; in which cafe, fome metallic flag, perhaps, might be ufeful.

SECT. LIV.

When fuch ores are to be reduced whofe metals are very apt to calcine, fuch as tin, zinc, &c. it might perhaps be of fervice to add fome phlo-gifton, fince the charcoal cannot afford enough of it in the open fire of thefe effays : fuch a phlogifton might be the hard refin, or fome fuch body. The manner of melting the volatile metals out of their ores *per defcenfum* might alfo, perhaps, be imitated : for inftance, a hole might be made in the charcoal, wide above, and very narrow at the bottom ; a little piece of the ore being then laid at the upper end of the hole, and covered with fome very fmall pieces of the charcoal, the flame muft be directed on the top : the metal might, perhaps, by this me-thod gather in the hole below, concealed from the violence of the fire, particularly if the ore is very fufible, &c.

Several of my experiments have indeed induced me to believe the poffibility of thefe improvements ; but as I have not yet had an opportunity of bring-ing them to perfection, 1 will not deliver them as infallible : thefe hints are only communicated as an inducement to farther experience,

SECT.

SECT. LV.

The ufe of the Pocket-Laboratory, as here defcribed, is chiefly calculated for a travelling mineralift. But a perfon who is always refiding at
one and the fame place, may by fome fmall alteration make it more commodious to himfelf, and
avoid the trouble of blowng with the mouth.
For this purpofe he may have the Blow-pipe
go through a hole in a table, and fixed underneath to a fmall pair of bellows with double bot-
toms, fuch as fome of the glafs-blowers ufe, and
then nothing more is required, than to move the
bellows with the feet during the experiment; but
in this cafe a lamp may be ufed inftead of a candle.
This method would be attended with a ftill greater
advantage, if there were many fuch parts as fig. 3.
tab. 1. the openings of which were of different
dimenfions: thefe parts might by means of a fcrew
be faftened to the main body of the Blow-pipe, and
taken away at leifure. The benefit of having thefe
nozzles, if I may be permitted to call them fo, of
different capacities at their ends, would be that of
exciting a ftronger or weaker heat as occafion
might require. It would only be neceffary to obferve, that in proportion as the opening of the pipe
(nozzle) is enlarged, the quantity of the flame
muft be augmented by a thicker wick in the lamp,
and the force of blowing increafed by means of
weights laid on the bellows. A much intenfer
heat would thus be procured by a pipe of a confiderable opening at the end, by which the experiments might undoubtedly be carried farther than
with the common Blow-pipe.

SECT.

SECT. LVI.

A traveller, who has feldom an opportunity of carrying many things along with him, may very well be contented with this Pocket-Laboratory, and its apparatus, which is fufficient for moft part of fuch experiments as can be made on a journey. There are, however, other things very ufeful to have at hand on a journey, which ought to make a fecond part of the Pocket-Laboratory, if the manner of travelling does not oppofe it : this confifts of a little box including the different acids, and one or two matraffes, in order to try the mineral bodies in liquid menftrua, if required.

SECT. LVII.

Thefe acids are, the Acid of Nitre, of Vitriol, and of Common Salt. Moft of the ftones and earths are attacked, at leaft in fome degree, by the acids ; but the calcareous are the eafieft of all to be diffolved by them, which is accounted for by their calcareous properties. The acid of nitre is that which is moft ufed in thefe experiments; it diffolves the limeftone, when pure, perfectly, with a violent effervefcence, and the folution becomes clear : when the limeftone enters into fome other body, it is neverthelefs difcovered by this acid, through a greater or lefs effervefcence in proportion to the quantity of the calcareous particles, unlefs thefe are fo few, as to be almoft concealed from the acid by the heterogeneous ones. In this manner, a calcareous body, which fometimes nearly refembles a filiceous or argillaceous one, may be known from thefe latter, without the help of the Blow-pipe, only by pouring one or two drops of this acid

upon

upon the fubject; which is very convenient when there is no opportunity, nor time, of ufing this inftrument.

SECT. LVIII.

The Gypfa, which confift of lime and the vitriolic acid, (Sect. xviii. 12.) are not in the leaft attacked by the acid of nitre, if they contain a fufficient quantity of their own acid, becaufe the vitriolic acid has a ftronger attraction to the lime, than the acid of nitre: but if the calcareous fubftance is not perfectly faturated with the acid of vitriol, then an effervefcence arifes with the acid of nitre, more or lefs in proportion to the want of the vitriolic acid. Thefe circumftances are often very effential in diftinguifhing the *calcarei* and *gypfa* from one another.

SECT. LIX.

The acid of nitre is likewife neceffary in trying the zeolites, of which fome fpecies have the fingular effect to diffolve with effervefcence in the abovementioned acid; and within a quarter of an hour, or even fometimes not until feveral hours after, to change the whole folution into a clear jelly, of fo firm a confiftence, that the glafs, wherein it is contained, may be reverfed, without its falling out.

SECT. LX.

If any mineral body is tried in this menftruum, and only a fmall quantity is fufpected to be diffolved, though it was impoffible to diftinguifh it with the eye during the folution, it can be eafily difcovered

by

by adding to it *ad faturitatem* a clear folution of an alcali, when the diffolved part will be precipitated, and fall to the bottom. For this purpofe the *fal fodæ* (Sect. xx.) may be very ufeful.

S E C T. LXI.

The acid of nitre will fuffice for making experiments upon ftones and earths; but if the experiments are to be extended to the metals, the other two acids (Sect. lvii.) are alfo neceffary. As the acids are very corrofive, they muft not be kept in the ordinary Pocket-Laboratory, already defcribed, for fear of fpoiling the other apparatus, if the ftoppers fhould happen not to fit exactly to the necks of the bottles, and fome of the acid fhould be fplit.

S E C T. LXII.

I have a feparate box, which is eight inches and three quarters long, four inches broad, and five inches high. In this box are three long and narrow bottles, containing the acids, placed upright at one end of it, two glafs matraffes laid horizontally in the upper part, and a little drawer underneath, made on purpofe to fill the empty room below the matraffes, and to give the box a regular form; and as charcoal is not every where to be met with in travelling, I always have a piece in this drawer for the ufe of the Blow-pipe.

S E C T. LXIII.

In order to keep the acids more clofe in the bottles, fince the glafs-ftopper is not always fufficient, I have a glafs-cover befides, made fo, as to fcrew

round

round the neck of the bottle; and if this is nicely made, nothing can come through, though the box be inclined, or even reverfed, which fometimes may happen. The natural form and bignefs of the glafs matraffes is feen tab. 1, fig. 4. They ought to be very thin at the bottom, that they may not crack, by being fuddenly put over the fire, or taken off it. In thefe matraffes folutions may very eafily be made over the flame of a candle : every mineral body capable of being affected by the acids in this degree of heat, may here be dif-folved, and particularly the metals. As the ma-nagement in thefe proceffes is the fame as in ordi-nary laboratories, of which we have ample defcrip-tions in feveral books, it is not neceffary to copy them here, my intention being only to defcribe an eafy way of making experiments upon mineral bo-dies, which has not before been publifhed; in ex-plaining of which I neverthelefs have been forced now and then to mention fomething that more properly belongs to Mineralogy.

SECT. LXIV.

Another inftrument is likewife neceffary to a complete Pocket-Laboratory, viz. a Wafhing-trough, in which the mineral bodies, and particu-larly the ores, may be feparated from each other, and from the adherent rock, by means of water.

This trough is very common in the laboratories, and is ufed of different fizes; but here only one is required of a moderate fize, fuch as twelve inches and a half long, three inches broad at the one end, and one inch and a half at the other end, floping down from the fides and the broad end to the bot-tom, where it is three quarters of an inch deep : I have given a figure of it in tab. 1. fig. 5. It is

com-

commonly made of wood, which ought to be chofen fmooth, hard and compact, wherein are no pores in which the minute grains of the pounded matter may conceal themfelves.

It is to be obferved, that if any fuch matter is to be wafhed, which is fufpected to contain fome native metal, as filver or gold ; a trough fhould be procured for this purpofe, of a very fhallow flope, becaufe the minute particles of the native metal have then more power to affemble together at the broad end, feparate from the other matter.

S E C T. LXV.

The management of this trough, or the manner of wafhing, which I fuppofe to be known before, confifts chiefly in this : That when the matter is mixed with about three or four times its quantity of water in the trough, this is kept very loofe between two fingers of the left hand, and fome light ftrokes given on its broad end with the right, that it may move backwards and forwards, by which means the heavieft particles affemble at the broad and upper end, from which the lighter ones are to be feparated by inclining the trough and pouring a little water on them. By repeating this procefs, all fuch particles as are of the fame gravity may be collected together, feparate from thofe of a different gravity, provided they all were before equally pounded ; though fuch as are of a clayifh nature, are often very difficult to feparate from the reft, which, however, is of no great confequence to a fkilful and experienced wafher. The wafhing procefs is very neceffary, as there are often rich ores, and even native metals, found concealed in earths and fand in fo minute particles, as not to be difcovered by any other means.

F I N I S.

AN
ALPHABETICAL TABLE

OF DIFFERENT

MINERALS,

WITH THE

SWEDISH and GERMAN Names added thereto,

For rendering more eafy the Knowledge of *Mineralogy*;
a Subject treated of by fome Writers, who make Ufe
of feveral original Terms taken from thofe Two Languages.

ACIDS, (mineral) *Sw.* Mineralifka Syror ; *Germ.* Saure-
Saltz ; *or* Mineralifche-Sauron. Section 120.
Acid of common falt, *Sw.* Kok-falt-fyra ; *Germ.* Koch-
falz-faure. Sect. 127. Vid. Vitriolic.
Agates. The irregular nodules in which they are found,
are called *Amorphi*, Page 69, in the note.
Argillacea (terra) Sw. Ler-arter ; *Germ.* Thon-arten. Sect. 77.
Alabaftrites, *Sw.* Strat-gips. Sect. 17.
Alcali, *(Sal) Sw.* Alcalifka.
Alcali of the Sea, *Sw.* Haf's Salt's Alkali. Sect. 136.
Alkaline mineral falts, *Germ.* Mineralifche-laugen Salter.
Sect. 135.
Alum, *Sw.* Alun ; *Germ.* Alaun.
Common Alum ore, *Sw.* Vanlig alum malm ; *Germ.* Ge-
mines alaun erts. Sect. 124. C.
Alum like Vitriol, *Germ.* Galitzenftein. Sect. 231, p. 219.
Alumen plumofum, Sw. Gedigen-alun ; *Germ.* Fieder-alaun,
or Gediegner-alaun. Sect. 124, B. 2. *a.*
Minera aluminis alba, Sw. Hvit-alun malm ; *Germ.* Weifes
alaun erts.

Y   Amber

Amber, *Sw.* Ambra.
   Amber Succinum, Sw. Bernften. Sect. 133.
Ammoniacum fixum Naturale *(Sal) Sw.* Salt-afka. Sect. 21.
Amygdaloides, *Sw.* Mandelften. Sect. 268.
Antimony, *Sw.* Spits glas ; *Germ.* Spies glas. Sect. 232.
   Crude antimony, *Sw.* Skierften. Sect. 237.
Aphronitrum, Sw. Mur-falt; *Germ.* Mauer-faltz. Sect. 137.
Arfenic, *Sw.* and *Germ.* Arfenik.
   Arfenic iron ore, *Sw.* Mifpickel. Sect. 114.
   Native Arfenic, *Germ.* Scherben cobolt, and Fliegenftein.
      Sect. 239.
   Native Arfenic with fhining fiffures, *Germ.* Spigel Cobolt.
      Sect. 239. C.
   Yellow Arfenic, *Sw.* Orpement ; *Germ.* Auripigment. Sect.
      241.
Realgar mineralis, Germ. Gediegen Raufchgelb. Sect. 241.
Arfenicum ferro fulphurato mineralifatum, Sw. Gift-kies ; *Germ.*
   Raufchgelb kies.
A fre of an imperfect kind, *Sw.* Galtfnas, Sad Slag, Brinda.
   See the Note at p. 82.
Afbeftus membranaceus, Sw. Berg Lader, and Berg Kiot;
   Germ. Berg Fleifch. Sect. 103. A.
Suber montanum, Sw. Berg Kork. Sect. 164. B.
Earth Flax, *Sw.* Berglin; *Germ.* Berg Flachs, Sect. 105. A.
Broken earth flax, *Sw.* Sad Slaglin. Sect. 106.

Bafaltes, *Sw.* Skiorl; *Germ.* Schirl ; Saulenftein. Sect. 72.
Sparry Bafaltes, *Sw.* Skiorl-fpat. Sect. 73.
   The Pin-ore Bafaltes, *Sw.* Gran-ris-malm. Sect. 74. *a.*
   Shirl with concentrated fibres, *Sw.* Stiern flag. Sect. 74. *b.*
Bifmuth, *Sw.* Afkbly, Vifmut ; *Germ.* Wiffmuth. Sect. 221.
   Flowers of Bifmuth, *Sw.* Vifmut blute, Sect. 223.
   Bifmuth mineralifed with fulphur, *Sw.* Vifmut glans. Sect.
      224. *n.* 1.
Blac-lead, *Sw.* Blyerts ; *Germ.* Bleyerts, Wafferbly. Sect. 154.
Blood-ftone. See Iron.
Bog-ores, *Sw.* Siômalm; *Germ.* Suerts and Sumpferts. Sect. 202.
Bole, *Sw.* Iern Lera ; *Germ.* Eifenthon. Sect. 85.
   Indurated Bole, *Sw.* Flets-malm. Sect. 87.
   Terra rubrica, Sw. Rod-krita; *Germ.* Rothel Kreida. Sect. 86.
   Slaty Bole. *Sw.* Skifver Lera. Sect. 87.
   Of Scaly Particles, *Sw.* Hornblende. Sect. 88.
Bononian ftone, *Sw.* Bononifk-fpat ; *Germ.* Bologuefer-fpat.
   Sect. 18. A.

Cæruleum montanum, Sw. Bergblott; *Germ.* Bergblau. Sect. 194.
Calcareous earths, *Sw.* and *Germ.* Kalk-arten. Sect. 4.
                           Phlogiftum

Pyrites cupri, Sw. Gul Köpper malm; *Germ.*Fahl Kupfererts. Sect. 198.

Liver brown copper ore, *Sw.* Kopper lazur. Sect. 198.

Crofs-ftone, or *Lapis cruciferus, Germ.* Bafler Taufstein, Sect. 75.

Cryftallus montana, Sw. Berg-kryftall. Sect. 52.

Rhomboidal cryftals, *Sw.* Moffgrufvan. Sect. 66.

Pebble criftal, *Sw.* Wattn-kyftall. Sect. 53, in the note.

Cruft, or Coat, *Sw.* Kaper.

The cruft which covers the agates, *Sw.* Agate-gall.

Decompounding away, or weathering, *Sw.* Wittring.

Diamond, *Sw.* Diamant. Sect 42.

Dufk, or Dark, *Sw.* Fluff. *ver. gr.* Amethift fiufs.

Earths in general, *Sw.* Jord-arter ; *Germ.* Erd-arten. Sect. 3,

Englifh earth, or *Cologne Clay* of Sect. 78, *Sw.* Engelfk-jord.

Emerald, *Sw.* Smaragd. Sect. 48.

Emery, *Sw.* Smergel ; *Germ.* Smirgel. Sect. 113. *n.* 2.

Flint, *Sw.* Kifel flinta ; *Germ.* Hornftein. Sect. 54.

Siliceous earth, *Sw.* Kifel-arter; *Germ.* Kiefel-arten. Sect. 40.

Common flint, *Sw.* Boffe-flinta ; *Germ.* Teverftein, Sect. 61.

Small Flints and pebbles for ballaft, *Sw.* Singel. Sect. 62.

Chert, *Sw.* Helleflinta ; *Germ.* Hornftein. Sect. 63.

Flos ferri, Sw. Eifchblute, and Eifenblute. See Iron.

Fluor of a polygonal form, *Sw.* Slag. Sect. 99. No. 13.

Fluors in general, *Sw.* and *Germ.* Flufs. Sect. 97.

Fullers earth, *Sw.* Walkfera. Sect. 84. *A.*

Fufible, *Sw.* Quicka.

Fufible rock, *Sw.* Quickften.

Garnet (coarfe) *Sw.* Granat-berg. Sect. 69. *A.* 1.

Glacies Mariæ, Sw. Marien-glafs. Sect. 18.

Gold, *Sw.* Guld.

Native gold, *Sw.* Gediget-guld ; *Germ.* Gewachfen gold. Sect. 165.

Leaved gold, *Sw.* Angefloget-guld. Sect. 165. *n.* 1.

Solid gold, *Sw.* Maffiot guld. Sect. 165. *n.* 2.

Gold in a cryftalline figure, *Sw.* Drufigt-guld. Sect. 165, *n.* 3.

Gold duft, *Sw.* Wafk-guld. Sect 165. *n* 4.

Gold mineralifed with mercury and filver, *Sw.* Guldisk-cinnober. Sect. 166. *c.*

Gold mineralifed with filver, zink, and iron, *Sw.* Schem-nizer-blende, *Germ* Kiegelerts.

Grains, *Sw.* Graupe.

Granite,

Granite, *Sw.* Graberg, and Graften. Sect, 270.
Loofe granite, *Sw.* Gintften. Sect. 270.
Gypfeous earth, *Sw.* Gur, or Kimmels-miol. Sect. 14.
Gypfum, *Sw.* Gips. Sect. 13.
Cryftallifed gypfum. *Sw.* Gips-drufer. Sect. 19.
Wedge-formed ditto, *Sw.* Gips-viggar ; *Germ.* Gips-keile, p. 26. *A.*
Stalactitical gypfum, *Sw.* Gips-finter ; *Germ.* Gipfa-niger, tropfftein. Sect. 20. See Plafter.

Hæmatites. See Iron-ore.
Holes of Drufen-fpars, *Sw.* Drake. Sect. 11, note.
Humus. See Mould.

Iceland Agate, *Sw.* Iflands agat ; *Germ.* Glas agat. Sect. 295.
Iron, *Sw.* Iern ; *Germ.* Eifen. Sect. 201.
Iron ore in grains, *Sw.* Smamalm. Sect. 202.
Ditto in lumps, *Sw.* Orkes. Sect. 202.
Scaly ditto, *Sw.* Iern glimmer; *Germ.* Eifenman. Sect. 203.
Cellular ditto, *Sw.* Kift formig. Sect 203. *e.* 3.
White ditto, *Germ.* Sthalftein, and Weifes Eifeners. Sect. 30.
Cellular calcareous iron ore, *Sw.* Kift formig. Sect. 33. 4, *b.* 2.
Iron fand ore, *Sw.* Iern fand malm.
White fparlike iron, *Sw.* Hvit Iern malm. Sect. 30.
Black iron ore, *Sw.* Svart Iern malm. Sect. 111. *b.*
Cold fhort iron, *Sw.* Kall breckt. Sect. 215.
Red Short, *Sw.* Rod breckt. Sect. 215.
Tough iron, *Sw* Enfuidt. Sect. 215.
Iron which wants no admixture, *Sw.* Sielf gangande. Sect. 215.
Flos ferri, Germ. Eifenblute. Sect. 33. *b. n.* 2.
Martial jafper, *Sw.* Iernig jafpis, Sect. 64.
Hæmatites, Germ. Blutftein. Sect. 203.
Fibrous hæmatites, *Sw.* Torrften.
Hæmatites nigrefcens, Sw. Svart glas kopf. Sect. 204.
Scaly red hæmatites, *Germ.* Eifenram. Sect. 205. *b.*
Blood-ftone, *Sw.* Blod ften. Sect. 203.
Red blood ftone, *Sw.* Rod glas kopf. Sect. 205.
Iron ore in form of folid balls, *Sw.* Purle malm. Sect. 202.
Ditto in form of porous balls, *Sw.* Skragg malm. Sect. 202.
Ditto in flat cakes, *Sw.* Penninge malm. Sect. 202.
Ochra ferri, Sw. Iern okra. Sect. 202. 1.
Bog ore, *Sw.* Siomalm ; *Germ.* Suertes, and Sumpferts. Sect. 202. 2.

Iflandicum fpatum. Vid. Spatum.

Lac

Lac Lunæ, *Sw*. Bleke. Sect. 5.
Lapis cruciferus. See Crofs-ftone.
Lapis ollaris, *Sw*. Telgften. Sect. 80, and 265.
Lapis fuillus,*Sw*.Orften ; *Germ*. Saufteni,and Stink-ftein. Sect.23.
Larvæ, or petrifications, *Sw*. Vandlingar ; *ver. gr. Terræ lar-vatæ,* Jord Vandlingar.
    In chalk, *Sw*. Kalk Vandlingar.
    Ditto in flint, *Sw*. Flint vandlingar, &c. Sect. 272, and 281.
Lead, *Sw*. Bly; *Germ*. Bley. Sect. 184.
    Cerufa nativa, *Sw*. Bly okra ; *Germ*. Bley ocher. Sect. 185,
    Lead Spar, *Sw*. Bly fpat.
    Blue Potter's ore, *Sw*. Bly glans, and Bly fcheweif ; *Germ*.
    Bley glantz, and Bley fcheweiff.
    Steel grained lead, *Sw*. Bly ftal malm. Sect. 188.
    Sparkling ditto, *Sw*, Skigg malm. Sect. 188.
    Lead ore with fulphurated iron and filver, *Sw*. Iernhaltig
    bly glans. Sect. 189.
    Ditto with antimony and fulphur, *Sw*. Strip malm ; *Germ.*
    Striputs.
Limeftone, *Sw*. Kalkften ; *Germ*. Kalkftein. Sect. 9.
    Coarfe grained ditto, *Sw*. Salt Slag. Sect. 8.
    White Limeftone, *Sw*. Kritften.
    Common Limeftone, *Sw*. Telgften, or Alfvarften, or
    Oelandften.
    Scaly Limeftone, *Sw*. Limften, and Limberg. Sect. 9.
    White and green, *Sw*. Storgrufvan.
    White and black, *Sw*. Herr ftens bottn.
Liver-ftone, *Sw*. Lefverften ; *Germ*. Leberftein. Sect. 18. *n.* 3.
Load-ftone, (coarfe and fcaly) *Sw*. Magnetiskt-eiffen-glim-
  · mer. Sect. 111. *d.*
Lufus naturæ, or cafual figures in minerals, *Sw*. Stengyckel.

Manganefe ore, or earth, *Sw*. Brunftens arter ; *Germ*. Braun-
    fteins-arten. Sect. 113.
Stony manganefe, *Sw*. Brunften. Sect. 113.
Marcafite. See Pyrites.
Marle, *Sw*. Mergel. Sect. 25.
    Semi-indurated Marle, *Sw*. Mergel skifver. Sect. 28.
    Stone marle, *Sw*. Malreka and Necrebrod ; *Germ*. Dukftein
    or Tophftein.
Marbles indurated, *Sw*. Malrekor.
Marmor metallicum, *Sw*. Tung-fpat. Sect. 18.
Marrow (ftone) *Sw*. Sten merg ; *Germ*. Stein mergel. Sect. 84.
    Stone marrow like foap, *Sw*. Keffekil. Sect. 84.
Martial earth (bricks made of) *Sw*. Water klinkert. Sect. 83.
Martial foap rock, *Sw*. Iern holtig fpecr'ften, Sect. 83.
                                 Metals.

Metals, *Sw.* Metaller ; *Germ.* Metalle.
Mica, *and* Micaceous, *Sw.* and *Germ.* Glimmer.
    Large plates of Mica, *Sw.* Chludna and Rufs glas ; *Germ.*
    Ruffifch glas.
    Small plates of mica, *Sw.* Kattfilfver ; *Germ.* Kazen filbet.
    Crumpled mica, *Sw.* Talc. Sect. 94.
    Mica Drufica, *Sw.* Talc/Drus. Sect. 95.
    Mica fquamofa martialis, *Sw.* Kattgull.
    Crumpled mica martialis, *Sw.* Wrefig-glimmer. Sect. 95.
Mineral, *Sw.* Mincraliska.
Mock-lead, *Sw.* Wolfram.
Mountain blue, *Sw.* Berg blot ; *Germ.* Berg-blau. Sect. 34.
Mountains, *Sw.* Fiell.
Mould, *Sw.* Mylla.
    Black mould, *Sw.* Mat jord ; *Germ.* Sumpfende, and Stan-
    berde. Sect. 293.
    Humus animalis, *Sw.* Diur jord ; *Germ.* Thiererde.
    Humus conchaceus, *Sw.* Sneck mylla.
    Humus lacuftris, *Sw.* Dy. Sect. 293. *b.* 2.
    Humus vegetabilis, *Sw.* Vext-mylla, and Vext-jord. See
    Turf.

Naphta, *Sw.* Berg-balfam. Sect. 148.
Native, *Sw.* Naturligt, and Gediget.
Nickel mineralifed, *Sw.* Kupfernickel. Sect. 252.

Ochre. See Iron.
Ollaris. Vid. Lapis Ollaris.
Ore, *Sw.* Malm.
Opal, *Sw.* Elementften. Sect. 55.

Pea-ftone, *Germ.* Sprudelftein. Sect. 12. 1. *a.* 1.
Pebbles (loofe) *Sw.* Kiflar
Pearl flag, *Sw.* Perle flag. Sect. 298.
Petrificata, *Sw.* Petrificater ; *Germ.* Verfeinerungen.
Petroleum, *Sw.* Berg-olia ; *Germ.* Berg-ol. Sect. 147.
Phlogifta mineralia, *Sw.* Jord fettmor ; *Germ.* Erd-harze.
    Sect. 144.
Pipe-clay. *Sw.* Pip-lera. Sect. 86. and 95.
Pitch rock oil, *Sw.* Berg tiera ; *Germ.* Berg thear. Sect. 149.
Pix montana, *Sw.* Berg beck ; *Germ.* Berg pech. Sect. 150.
Plafter ftone (common) *Sw.* Gips-ften. Sect. 16.
Loofe plafter, or in powder, *Sw.* Gur, *or* Himmels-miol.
    See Gypfum.
Plum-pudding ftone, *Sw.* Kifel breccia. Sect. 273.
Porcelana (*terra*) *Germ.* Porcelain thon, *and* Feuer-beftandi-
    ger-thon. Sect. 78.

                       Porphyry,

Porphyry, *Sw.* Porphyr. Sect. 266.
Powder, [impalpable flime] *Sw.* Slamm. Sect. 5. and p. 14.
 N. *.
Pumex, Sw. Pims-ften; *Germ.* Bims-ftein. Sect. 297.
Pyrites, *Sw.* and *Germ.* Kies.
 Pale yellow ditto, *Sw.* Blekel-gul-fvafvel kies. Sect. 152.
 Liver-coloured ditto, *Sw.* Wattn kies; *or,* Tenbett
  lefver flag. Sect. 153.
Copper and iron marcafite, *Germ.* Kupfer kies. Sect. 155.

Quartz, *Sw.* Katt flinta, and Hvit flinta. Sect. 50.
Rhombic quartz, *Sw.* Felt fpat. Sect. 66.
Iron ore with quartz, *Sw.* Torrften. Sect. 77. Note.
Grained quartz, *Sw.* Torrquartz.
Quick-filver, *Sw.* Queck-filber.

Red, *Sw.* Rod.
Refractory, *Sw.* Torra.
Regulus, Sw. Skierften.
 Rich copper Regulus, *Sw.* Trotften.
Rhenifh mill-ftone, *Sw.* Renlandsk Quarnften. Sect. 298.
Rock-ftone, *Sw.* Stellften.
 A kind of Rocks fpoken of in Sect. 39. *Sw.* Graberg.
Rocks, (high) *Sw.* Fiell.
Ruble-ftone, *Sw.* Gorften. Sect. 39.
Rubrica,(terra) Sw. Rod krita ; *Germ.* Rothel kreide. Sect. 86.
 a. 3.
Ruby, *Sw.* Rubin. Sect. 43.

Salt, *Germ.* Salz.
 Sal fontanum, Sw Kel-falt; *Germ* Brunnen-falz. Sect. 131.
 Sal urinofum, Sw. Rot-falt. Sect. 140.
 Sal marinum, Sw. Haf's-falt; *Germ.* Meer falz. Sect. 130.
 Sal montanum, Sw. Berg-falt. Sect. 129. *B.*
 Sal commune, Sw. Kok-falt.
 Sal ammoniacum fixum, Sw. Salt-aska; *Germ.* Salz-afche.
  Sect. 128.
 Sal ammoniacum naturale, Sw. Salmiac. Sect. 132.
 Acid Salts, *Sw.* Sura falter, Sect. 120.
 Salia neutra, Sw. Medel-falter; *Germ.* Mittel-falze.
 Weak falt-water, *Sw.* Salen. Sect. 131.
Sand-ftones, *Sw.* Sandften. Sect. 276.
Saxa compofita, Sw. Sammen-fatte, and Stelle-arter. Sect.
 260.
 Saxum compofitum ex mica et hornblende, Sw. Gronften.
 Saxum compofitum ex variorum fragmentis, Sw. Stelearts
  breccia,

                Saxum

Saxa conglutinata, *Sw*. Samman gyttrade Stelle arter. Sect. 260. *n*. 3.

Saxum compofitum micâ et quartz, *Sw*. Stelflén. Sect 262.

Ditto, *ex micâ, quartz & granato*, *Sw*. Norrka, and Murkften. Sect. 263.

Ditto, *ex jafpide martiali molli, feu argilli*, &c. *Sw*. Trapp-fkiol, *or* Tegel-fkiol, *or* Swart-fkiol, *or* Blabeft ; *Germ.* Schwach, *or* Schwarts ftein.

Scoriæ Vulcanorum, *Sw*. Naturliga flager ; *Germ.* Naturliche Schlaken. Sect. 294.

Serpentine ftone, *Germ*. Serpentin ftein. Sect. 81.

Selenites, *Sw*. Gips-fpar. Sect. 18.

Semi-metals, *Sw*. Halfwa-metaller ; *Germ*. Halb metalle.

Shirl, *Sw*. Skiorl, *and fometimes* Wolfram, Note p. 125 ; *Germ*. Saulenftein. Sect. 72.

Coarfe fhirl, *Sw*. Skiorl berg.

Striated fhirl, *Sw*. Strat fkiorl. Sect. 74.

Starred fhirl, *Sw*. Stiern-flag.

Siliccous. Vid. Flint.

Silver, *Sw*. Silfver ; *Germ*. Silber.

Pure native filver. *Sw*. Berg flint filfver.

Capillary filver, *Sw*. Har filfver.

Glafs filver ore, *Sw*. Silfver glas ; *Germ*. Glaferts. Sect. 169.

The goofe dung ore, *Sw*. Ganskotig erts. Sect. 168.

Red filver ore. *Sw*. and *Germ*. Rot gulden. Sect. 170.

Grey filver ore with antimony, *Germ*. Leberertz. Sect. 73.

Horn filver ore, *Sw*. Hornertz. Sect. 177.

Silver ore with fulphur, lead. and antimony, *Sw*. Strip malm ; *Germ*. Striperz. Sect. 176.

With fulphur and lead. or Galena, *Sw*. Bly glans ; *Germ*. Bleygantz. See Lead.

Silver ore in balls, *Sw*. Kugel erts. Sect. 175.

With copper, fulphur, and zinc, *Sw*. Beck blend ; *Germ*. Pech blende. Sect. 175.

With copper, antimony, and fulphur, *Sw*. Dalfalertz. Sect. 174.

Plumofe filver, *Germ*. Feder-ertz. Sect. 163.

White filver ore, *Germ*. Weiffertz. Sect. 172.

Black or footy filver ore, *Germ*. Ruffigtes ertz, and Silber fchwartz. Sect. 171.

Grey filver ore, *Sw*. Falerts, *and* Gra malm. Sect. 171.

Slate, *Sw*. Skifver.

Z.

Table

Table flate, *Sw.* Taffel fkifver ; *Germ.*. Talc fkifver. Sect. 264.

Smectis Briançon, *Sw.* Brianzoner krita. Sect. 79.

Soap rock, *Sw.* Speck ftén. Sect. 79.

Spar, *Sw.* Spat.

    Double refracting fpar, or *Spatum iflandicum,* *Sw.* Dubbel ftén ; *Germ.* Doppel ftein. Sect. 10.

    Stalactitical ditto, *Sw.* Droppften, and Skorpften ; *Germ.* Tropfftein, and Rindenftein. Sect. 12.

    Calcareous fpar, *Sw.* Kalk fpat.

    Fluor fpatofus, *Sw.* Lys-fpat. Sect. 99.

    Common fpar, *Sw.* Ratfigtig fpat. Sect. 10. *a.* 2.

    Dog's teeth fpar, *Sw.* Swin-tender. Sect. 11. *b.*

    Cockle fpar, *Sw.* Skiorl fpat.

    Lead fpar, *Sw.* Bly-fpat.

    Balls of cryftallifed fpar, *Sw.* Spat-klot. Sect. 11.

    Holes filled with drufen fpar, *Sw.* Drake-hol. Sect. 11.

    Cryftallifed fpar, *Sw.* Kalk fpat drufer.

    Selenites, *Sw.* Gips fpat.

    Flux fpar, *Sw.* Flufs fpat.

Spuma Lupi, *Sw.* Wolfran. See Sect. 117.

Steel (like fteel) *Sw.* Stal flag.

Stone, *Sw.* Sten ; *Germ.* Stein.

    Touch-ftone, *Sw.* Prober ften. Sect. 250.

    Whet-ftone, *Sw.* Hwethelten, and Brynften. Sect. 264.

    Veins of ftony matter, *Sw.* Gangar, and Klyfter. Sect. 12. Note.

    Heap of ftones, Mundik, &c. *Sw.* Warp.

Succinum, *Sw.* Bernften ; *Germ.* Bernftein. Sect. 146.

Sulphur, *Sw.* Svafvel ; *Germ.* Shwefel. Sect. 125, and 151.

Superficial, *Sw.* Angeflogen.

Swine fpars, *Sw.* Orlten. Sect. 157.

Talc, *Sw.* Strat gips. Sect. 17.

Talc cubes. *Sw.* Talc-terringer. Sect. 96. See Mica

Terra, *Sw.* Arten ; *Germ.* Arter.

    Terra Tripolitana, *Sw.* Trippel. Sect. 89.

Tin, *Sw.* Tenn ; *Germ.* Zinn. Sect. 180.

    Tin ftone, *Sw.* Tenn berg ; *Germ.* Zinn ftein. Sect. 181.

    Tin grains, *Sw.* and *Germ.* Zinn-graupe. Sect. 181—84.

Topaz, *Sw.* Topas. Sect. 45.

    Smoaked topaz, *Sw.* Rok topas ; *Germ.* Rauch topas. Sect. 52.

Turf,

Turf, *Sw.* Torf; *Germ.* Dammerde torf. Sect. 293.
 Solid turf, *Sw.* Beck torf.
 Lamellated ditto, *Sw.* Papers turf. See Mould.

Veins (small) separating the load from the rock, *Sw.* Slepp-
 skiok, *or* Skioler.
Viride montanum, *Sw.* Berg gron : *Germ.* Berg-griin. Sect. 194.
Vitriol, *Sw.* Viktril.
 Blue vitriol, *Sw.* Bla viktril; *Germ.* Blauer vitriol. Sect. 122.
 Copper vitriol, *Germ.* Kupfer vitriol.
 Vitriol of zink, *Sw.* Hvit viktril; *Germ.* Weifen vitriol,
 and Galizen stein. Sect. 122 *n.* 3.
 Ditto of iron and copper, *Sw.* Salzber viktril. Sect. 123.
 Ditto of iron, zinc, and copper, *Sw.* Faln viktril. .
 Ditto of nickel and iron, *Sw.* Nikel viktril.
Vitriolic, Viktriller.
 Vitriolic acid, *Sw.* Viktrils-fyra ; *Germ.* Vitriolifke-faure.
 Sect. 121.
Volatile, *Sw.* Flygtigl ; *Germ.* Fluchliges.

White, *Sw.* Weifs.

Zinc, *Sw.* Spiauter.
 Fine sparkling zinc, *Sw.* Braun bley erts gofslaria. Sect.
 229. 3.
 Zinc mineralised with sulphurated iron, *Sw,* Blende. Sect.
 229.
 Lapis calaminaris, *Sw.* Gal meya. Sect 228. *a.* 1.

N. B. The POCKET LABORATORIES for Mineralogical
 Experiments. mentioned in the Note at Page 283, are still
 to be had at the same general Office, lately removed from
 Wood-ftreet, *Cheapfide*, to the Shop of Mr. *W. Brown*, Book-
 feller, at the Corner of *Effex-Street*, in the *Strand*, near
 Temple-Bar, London. .